SHIPPING IN THE BALTIC REGION

T0231632

Shipping in the Baltic Region

Edited by
MICHAEL ROE
Centre for International Shipping and Transport
University of Plymouth

LONDON AND NEW YORK

First published 1997 by Ashgate Publishing

Published 2017 by Routledge
2 Park Square, Milton Park, Abingdon, Oxfordshire OX14 4RN
711 Third Avenue, New York, NY 10017, USA

First issued in paperback 2017

Routledge is an imprint of the Taylor & Francis Group, an informa business

Copyright © M. Roe 1997

British Library Cataloguing in Publication Data

Shipping in the Baltic region. - (Plymouth studies in
    contemporary shipping)
    1. Shipping - Baltic Sea
    I. Roe, Michael, 1954-
    387.5'0916334

Library of Congress Catalog Card Number: 96-086731

ISBN 13: 978-1-138-26912-5 (pbk)
ISBN 13: 978-1-85972-501-6 (hbk)

Contents

Introduction

Michael Roe
Centre for International Shipping and Transport
University of Plymouth

This volume represents the first covering a series of research areas from a number of maritime sectors and topics. It is intended to provide the basis for a series of publications over the coming years which will concentrate upon the full range of maritime related disciplines in the widest sense, deriving from research undertaken primarily at the Institute of Marine Studies, University of Plymouth, but also collaborative ventures with other institutes around the world.

This specific text is the first in a series to be produced by the Centre for International Shipping and Transport at the University of Plymouth, UK, in collaboration with the Institute for Maritime Transport and Seaborne Trade at the University of Gdansk, Poland, dealing with a range of topics and issues central to the maritime sector and its associated industries.

There has been collaboration between the two institutes for many years, dating back to the period immediately before the major social, political and economic changes of the 1980s, and this has involved a mixture of collaborative research projects, staff exchanges and joint publications. Funding for this research has come from a variety of sources including the UK and Polish governments, the European Union and various UK and Polish industrial concerns.

The Centre for International Shipping and Transport is the UK's largest academic teaching and research centre for maritime studies, transport and logistics, and has interests across the spectrum of the discipline, but particularly specialising in economics, policy, politics, planning, law and finance in the ports, shipping, transport and logistics industries. Along with thriving undergraduate and postgraduate teaching programmes, there is also a substantial doctoral scheme which has involved collaboration with the University of Gdansk in projects concerning the development of the Polish liner sector and the relationship between flexible and fixed infrastructure with particular interest in the provision of shipping and port facilities.

The Institute for Maritime Transport and Seaborne Trade at the University of Gdansk is the premier academic institute in the former Eastern Europe specialising in a range of maritime studies in the Baltic region, having developed these interests throughout its growth from the Communist government days through to the new environment. It has a long standing reputation in maritime research and a series of thriving postgraduate and undergraduate schemes in the maritime, transport and logistics fields with particular interests in the Polish liner sector - a major international industry and locally centred in Gdynia - and in the changes in the maritime industries that have taken place since the decline of the old regimes of Eastern Europe and the attempts being made to meet the future needs of the region, sector and industry.

Ludwig Kondratowicz, Konrad Misztal, Zofia Sawiczewska, Stanislaw Szwankowski, Leopold Kuzma and Janusz Zurek are all associated with the Institute of Maritime Transport and Seaborne Trade at the University of Gdansk whilst each have specific interests in the maritime sector of Poland, including in many cases roles within the industry itself. This brings a valuable insight into the various sectors which is unavailable from Western experts or even many of those in Poland in other academic institutions.

Neal Toy, Gillian Ledger and Michael Roe are all attached to the Institute of Marine Studies at the University of Plymouth, and are actively involved in research in the Polish Baltic region. Neal Toy is currently examining infrastructural relationships in Poland in the maritime sector and has spent periods of time with both academic institutions and the industry in Szczecin, Gdynia and Gdansk. Gillian Ledger has recently completed a doctoral study of the adaptation of Polish liner shipping to the new political, economic and social environment and is now working as a Post Doctoral Research Assistant at the Institute of Marine Studies. Michael Roe is Professor of Maritime Policy at the Institute of Marine Studies, where his research activities concentrate upon the Baltic Region and the countries of developing central Europe.

The paper by Ledger and Roe examines the changing relationship between the European Union and the countries of Eastern Europe, in the field of shipping policy. It outlines and discusses the slowly emerging details of the shipping policy of the European Union including Stages One and Two and the more recent suggestions emanating from the Commission. Whilst doing so it emphasises the impact upon East Europe and the influence that East European policy makers have had upon Union policy. The significance of the Single European Market and new entrants are also discussed. Finally, the impact of change is East Europe is noted upon the future development of East-West policy making.

The paper by Kondratowicz presents a unique and highly topical view of the ferry feeder sector and the development of a new model approach to its interpretation and prediction. More specifically it presents a multimodal transport orientated simulation modelling system of a simple example of a simulated transport network that involves one ocean shipping line, a short sea link, road transport and two seaports. The work concentrates on the development of the model and presents detailed results of its application to scenarios in the Polish sector.

Misztal's paper examines the development of Polish ferry ports in the Baltic Sea region and in particular the specific facilities available in each, and the new proposals put forward for their improvement, concentrating upon the provision of new road facilities. Sawiczewska takes a long and very detailed look at the role and importance of short sea shipping based in Poland for the Polish economy and its significance in the international market place, providing recommendations for the future. Meanwhile, Szwankowski looks at a particularly interesting and topical new development in the region, in the form of fast ferries and their physical requirements in the seaports of the region. He concludes that the introduction of new technical facilities of this sort is likely in the market environment of the Baltic region and that the Polish ports will have a major part to play.

Toy and Roe's paper attempts to illustrate the major effects that East European economic and political changes have had upon a particular sector of Poland's shipping industry - the passenger ferry sector. The broad approach is to outline the structure, conduct and performance of the ferry industry during Communist rule, repeat this for the current situation, and finally account for the variations observed, if any are present. Many of the transitions relate to political, legal and structural entities, and as such, by their very nature, are more suitable for qualitative rather than quantitative appraisal. Analysis and conclusions are based upon the use of free response interviews with industry and academic authorities, secondary quantitative data and published literature.

Zurek discusses the new North-South motorway and its impact upon Polish ferries. This paper looks at the history of Polish motorway construction and the current plans for development in the coming years, with particular reference to the relationship between the A1 motorway and maritime investments.

Finally Kuzma's paper examines the restructuring and privatising of Poland's ports starting from the subjective - organisational analysis of the situation prior to the social, political and economic changes that have occurred and going on to examine the impacts subsequent to the developments.

Overall, it is hoped that this volume can add to the developing body of knowledge on the maritime sector of the region, drawing, as it does, in particular upon the experience of those directly involved both spatially and industrially. It would not have been possible without the continued collaboration of all our colleagues in Gdansk, in particular Janusz Zurek who has played a major role in developing relationships with the University of Plymouth over many years. Great thanks must also go to Gillian Ledger, who provided considerable editorial help, and last, but not least, to Marie Bendell at the Institute of Marine Studies for keeping me both on track and relatively sane when faced with increasingly sizeable word processing disasters.

However, the greatest personal thanks of all must go to Liz, Joseph and Siân for their endless patience and entertainment, and additionally to the Green Army, for a fabulous day out at Wembley in May 1996.

European Union shipping policy and Eastern Europe

Gillian Ledger and Michael Roe
Centre for International Shipping and Transport
University of Plymouth

Abstract

This paper examines the changing relationship between the European Union (EU) and the countries of East Europe, in the field of shipping policy concentrating only upon the period up to around 1992 when policies had to change dramatically to incorporate the transition in European politics and relationships that was occurring. It outlines and discusses the slowly emerging details of EU shipping policy, including Stages One and Two and a number of other suggestions emanating from the Commission. Whilst doing so, it emphasises the impact on East Europe and the influence that East European states have had upon Union policy.

The significance of the Single European Market and new community entrants are also discussed.

Finally, the impact of changes in East Europe is noted upon the future development of East-West policy making.

Introduction

With the changes taking place in East Europe, the attitudes of the European Union - hereafter referred to as the European Community or EC - towards the East have altered considerably. Merritt (1991) commented that the EC nations' policies towards Eastern Europe might be described as "chaotic and unpredictable", and that establishing a durable and consistent relationship between the EC and East Europe is essential although it will not be easy. This paper aims to examine the development of a Community shipping policy in the light of these changing attitudes and relationships, and to look at the impacts upon East Europe.

When the Community was founded in the late 1950s, the Soviet Union saw it as a reinforcement of the capitalist camp, detrimental to Soviet interests, and having decided not to accord it juridical recognition continued to treat it for over two more decades with varying degrees of coldness. The Community for its part paid only marginal attention to East Europe throughout the 1960s although as détente began to replace the cold war, EC members sought to outdo each other both economically and politically in their relations with the East. At first the British set the pace with a fifteen year credit for the Soviet Union in 1964. But the most important of this series of agreements was the Moscow Treaty between the Soviet Union and the Federal Republic of Germany, signed in 1970 which led to more normal relations between the two parties (Pinder 1991). Meanwhile the Treaty of Rome signed on 25th March 1957, under which the EEC was formed, had made no special provision for relations with the states of East Europe, apart from the 'Protocol on German international trade and connected problems'.

Transport and Shipping policy of the EC

According to Arbuthnott and Edwards (1979) the general aims of the Treaty of Rome included :

1   Drawing the people of Europe closer together.
2   Encouraging economic growth.
3   Improving living and working conditions.

The treaty specified three areas as being essential to unification of the Community; these were:-

1   Agriculture.
2   Social Services.
3   Transport.

As far as transport was concerned, the Treaty stated that:

The transport market must be organised in accordance with a market economy. Public intervention should occur only where it is otherwise impossible to proceed. The Community must ensure that restrictions to freedom to provide services are removed. At the same time the aim is to harmonise the overall framework in which modes and companies operate. Therefore we must not lose sight of the objective of optimising the transport process with a view to increasing competitiveness of the EEC, and improving services to the public (Treaty of Rome 1957).

Article 84(1) of the Treaty excluded sea and air transport from the common transport policy, stating that "the provisions of this title (i.e. Title IV, Transport, of Part Two, "Foundations of the Community") shall apply to transport by rail, road and inland waterway". Article 84(2), however adds:

> The council may, acting unanimously, decide whether, to what extent and by what procedure, appropriate provisions may be laid down for sea and air transport (Seefeld 1977).

Hence, for many years, the EC did not have a shipping policy as the need for one was considered to be excluded by the Treaty. Also the law making process of the EC was bureaucratically slow and shipping was not as prominent an issue with either governments or the electorate as numerous others such as agriculture or industrial development. The result was that between 1957 and 1977 there was no shipping legislation. The first moves were made when the UK and Ireland, both major seafaring nations, joined the EC in 1973, and in particular when the UK took the European Council of Ministers to the European Court of Justice over a narrow maritime issue, relating to the employment mobility of European seafarers, and won. During the 1980s, the EC registered fleet declined drastically compared to world fleets, and the need for a stated policy became more than evident (Commission for the European Communities 1989).

Between 1977 and 1985, in particular, several different shipping policy areas were investigated and developed, although little legislation was agreed. These policy areas can be summarised as follows:-

1   Social issues relating in particular to free movement of people - including seafarers' employment - between countries, harmonisation of seafarers' working conditions, and a range of social security measures relating to the maritime sector, for example social and health benefits.

2   The right to establish shipping companies throughout the EC; the Treaty made it illegal for countries to show bias against any non-national attempting to set up a shipping company. This related to

the issues of harmonisation and liberalisation.

3   Liner shipping conferences were recognised as violating the Treaty of Rome as they were anti-competitive. However, they did present strong political arguments for their retention and hence no progress was made to 1985.

4   Safety and pollution legislation in relation to shipping was agreed before 1985. The EC entered into negotiations with organisations such as BIMCO, and between 1978 and 1982, produced shipping related Directives, including those relating to pilots and tanker operations.

5   The "Brussels package" agreed by all member states in 1979 stated that ratification of the UNCTAD code by EC members was a requirement. The 40/40/20 agreement was to apply to Member States' trade with developing countries. Couper (1977) pointed out however that the code did not cover the activities of outsiders who do not belong to the conferences and hence would not affect operations of Comecon countries.

6   The Commission also recognised that the EC shipbuilding industry was facing economic difficulties and accepted that member states might have to provide subsidies. Although the EC has consistently been anti-subsidy in policy, a fixed but reducing percentage of ship cost (11% in 1991) was allowed to be subsidised. There is some evidence to show that these directives have been broken continuously since introduced.

7   In 1977 the European and Social Committee for the European Communities (ESCEC) produced an opinion on the "EEC's transport problems with East European countries". The report pointed out that the steadily mounting competition from the Eastern bloc in the field of maritime shipping was a cause for grave concern on account of the conditions under which it was flourishing. It added that because they were able to operate freely in the west, Eastern bloc countries were succeeding to an increasing degree in changing the pattern of East-West goods traffic in their own favour. Accordingly, the ESCEC pointed out that not only may Eastern bloc carriers' penetration of the markets threaten employment in transport, but in the long run, there may be grave drawbacks for industry in the Community as a whole. For this reason the Committee called on all the institutions

responsible for East-West transport questions to tackle this matter with the utmost vigour in order to ward off developments that would be disastrous for the economy and have grave social consequences (ESCEC 1977). The Committee then outlined the objectives they considered important in the specific field of sea transport as follows:

*   Comecon countries should be made to drop freight rates that are in no balanced relation to the normal terms in Western countries.

*   Community shipowners should be given a balanced share of bilateral traffic between Community and Comecon ports in both directions, at adequate rates and without carriers from other countries being excluded.

*   West European shipowners should be given the chance to acquire a share of traffic between Comecon ports and ports outside the Community.

*   Eastern bloc shipping lines should be allowed to accede to existing agreements between Western shipowners.

The ESCEC along with the Commission and the European Parliament had shown an interest in Eastern bloc activities for a number of reasons as it had created problems for EC shipping since the end of world war II. From 1977 onwards, the Seefeld Reports (Seefeld 1977 and 1979) investigated the area and reported back to the European Parliament and Commission. Eastern bloc shipping at this time was described as having a "conventional purpose of serving its own markets" and as performing "dubious defence exercises". All vessels were thought to have a defence bias or alternative defence purpose.

The first Seefeld report (23/3/77) called for the Commission to establish a common position toward state trading countries and other countries that wanted to build up their own merchant fleets. It suggested that pressure from outside the Community was forcing Europe to act, and highlighted the fact that state trading countries were "forcing their way onto the world shipping market and endangering the member states 'shipping industries". Until this time the Community had ignored these countries which in default of a clear cut Community decision on maritime policy were able to insist that all their export transactions were effected on a CIF basis and all their import transactions on an FOB basis, which left sea transport completely in their hands and eliminated European shipowners from this trade.

Even more important to the shipping industries of the Community countries was the threat represented by the widespread practice adopted by the state trading countries of undercutting conference tariffs in transport operations which linked directly to the member states but more especially in cross trades. Many western shipping companies were convinced that undercutting to an extremely low level in this way corresponded to dumping and pointed out that by western standards the Eastern bloc fleets were operating uneconomically in that, for example, vessel replacement costs were not being covered, although many of these arguments were disputed by a number of commentators including Bergstrand and Doganis (1987).

These points had already been identified in 1976 by Prescott (1976) who also noted the problematic practices of inter-governmental (either bilateral or multilateral) agreements reserving part or all of the cargo moving in the trade, and the establishment of joint shipping agencies in foreign countries without reciprocity; Western shipping businesses were not afforded the same possibility to run their business in Eastern bloc countries with the same freedom (Russell 1975).

The Seefeld report carried on to point out that state trading countries were "double-dealing", for example in connection with the code of conduct for liner conferences. On the one hand they voted with the developing countries in UNCTAD when it was a question of "thwarting the interests of the traditional maritime nations", on the other hand they adopted the position of an outsider by undercutting both industrialized and developing countries' shipping companies.

Seefeld suggested that if the Community did not develop a coherent sea transport policy, the goods to be shipped, would always be regarded as more important than the interests of those shipping them. Although the report did not suggest any policy alternatives, it did support the call from the shipping companies' associations for effective protection against the non-commercial practices of the state trading countries and hoped that the Community would raise shipping questions in any negotiations between the EEC and Comecon. Couper (1977) went on to suggest that the initial aims of a Community shipping policy should include negotiating as a Community with Comecon for the preservation of a co-ordinated seaborne trade.

The second Seefeld report (5/1/79) revealed that the EC was under pressure of time in working out and implementing a common transport policy from Comecon countries which were "anxious to push their way on to the world market with its rich pickings in foreign currencies" (Seefeld 1979).

The advance of industrialisation in the Comecon countries was seen as the reason for their emergence on world markets. The State trading countries penetrated world transport markets using all the resources at their disposal - for example at the cost of the standard of living of their citizens and without regard for profitability. As most Eastern bloc shipping companies were state owned they did not need to operate commercially, only to earn hard convertible, currency without which they would not be able to purchase vital foreign produce such as grain from the USA. Being able to charge less than cost because of state subsidy, these vessels were able to undercut western shipping. Although this was seen as unfair trading by the EC, little action could be taken, as EC countries had no jurisdiction over Eastern Europe, and could not risk taking action which might provoke retaliatory measures from East European countries. The EC did however set up a review of activities in the international trading and shipping industries of the CMEA with which it also entered into negotiations.

The Seefeld reports suggested that the EC needed to find ways to persuade the state trading countries to adopt a more western approach without destroying the foundations of free competition or forfeiting the high degree of efficiency that had characterised world sea transport.

Overall, until 1985 there was little shipping policy, and in particular legislation, to show for 28 years of the EC, although it is interesting to note the significance attached to East Europe throughout this period. The 1970s saw the beginnings of slow talks between the Comecon and the EC, and by the mid 1980s, the EC's relationship with the Comecon countries seemed to have stabilised at a modest level of activity. In June 1984 the Comecon summit meeting expressed its interest in a relationship with the Community, and in October of the same year a communication from Comecon to the Community suggested negotiations for an agreement, declaration or other document (Pinder 1991). By the time the first real EC shipping legislation was developed changing circumstances had caused East Europe to decline in its direct significance to the EC as a maritime competitor, and other issues became dominant.

Stage One : shipping policy

On 16th December 1986 the European Community, at a meeting of the Ministers of Transport, agreed a maritime package which combined with measures adopted earlier and outlined above, formed the basis of a Community Shipping Policy. The package included four regulations outlined by Erdmenger and Stasinopoulos (1988) as follows :

1. Council Regulation (EEC) No. 4055/86.
   Freedom to Provide Services.

This introduced the principle of freedom to provide services to intra community trade, distinguishing existing arrangements from future agreements. Applying to nationals of Member States, it aimed to prevent any Member State from discriminating in favour of its own shipping companies in another Member State.

It called for unilateral cargo restrictions by Member States to be phased out by 1st January 1993. Other discriminatory cargo sharing arrangements between Member States should be phased out or adjusted to comply with Community legislation. This regulation was of note to East European countries, as the

Council could extend it to nationals of a third country who provide shipping services and are established in the Community, for example Polish Ocean Lines' activities in the UK.

2. Council Regulation (EEC) No. 4056/86.
Competition Rules.

This aimed to apply Treaty competition rules to shipping and affected all international shipping services to and from Community ports, except tramp services.

It exempted liner cargo conferences from Treaty provisions on restrictive practices. Restrictive practices of Member States may prompt action from the Commission, although Council authorisation was required to deal with conflicts in international law. This had little, if any impact for East Europe as the shipping companies of these countries were usually non conference members.

3. Council Regulation (EEC) No.4057/86.
Unfair Pricing Practices.

This regulation applied to liner trades, and enabled a compensatory duty to be imposed on non EC shipowners by the Community if the following conditions were cumulatively present:-

i.   There are unfair pricing practices - defined as undercutting Community shipping services, where this is made possible because the non EC shipowners enjoy commercial advantages such as subsidy.

ii.   They cause financial or commercial injury to Community Members.

iii.   The interests of the Community make intervention necessary.

This regulation was particularly relevant to East European shipping companies who were believed to be undercutting western companies by charging less than cost because of state subsidy (Couper 1977).

4. Council Regulation (EEC) No 4058/86.
Co-ordinated action.

This regulation dealt with distortion of competition by governments giving preferential treatment to their own shipowners, and provided for co-ordinated Community action where third countries restrict access of EC shipping companies to ocean trades.

It was based on the assumption that it is in the interests of Community shipping not to encourage a protectionist approach in the rest of the world. The Community aimed to continue a commercial regime by taking action against non commercial attacks upon it. It is possible that this may have had some East European impact, especially where East European shipping companies had their offices in the EC rather than in East Europe.

Although Stage One was somewhat limited in scope it nevertheless provided the first legislative basis for an EC shipping policy. By the time Stage Two of a shipping policy for the EC emerged changes in East Europe such as the accession of Gorbachev, decreasing Soviet intervention in other East European states and declining East European economies meant that shipping in East Europe was no longer of such significance to the EC.

Stage Two : shipping policy

On 5th June 1989 the Commission produced two separate documents which constituted the basis for Stage Two of the EC shipping policy, which it was intended should be translated into legislation.

Paper 1. "A Future for the Community Shipping Industry: Measures to Improve the Operating Conditions of Community Shipping".
(Commission of the European Communities 1989).

Several policy initiatives were proposed in this paper:

1.   A Community ship register : EUROS.

This was to be introduced on the 1st January 1991, for vessels of less than 20 years, and greater than 500 grt. Run in parallel with national flags, EUROS would offer benefits such as:

-     Easier movement of ships between member states i.e technical compatibility.
-     Mutual recognition of seafarers' qualifications.
-     The 'opening up' of cabotage to all vessels on the Community register.

Vessels would be required to meet safety and certification levels of every Member State; additionally all officers plus half the crew had to be Community nationals. The aims of the scheme included the revitalisation of the EC national fleets.

Although it was expected that EUROS might come into effect as early as June 1991 (Lloyds List 16/1/91), disagreements over the issue of tax relief led to delays. On 19th November 1991, Lloyds List reported that the European Commission would adopt fresh proposals for a European ship

13

register by the end of the year pending final agreement on the new income tax relief measures for seafarers. By December 18th 1991, European Transport Commissioner Karel Van Miert said his latest amendments to the Commission's proposed EUROS ship register, giving income tax concessions to seamen on EUROS registered ships, had been welcomed by a majority of transport ministers. By 1996, despite protracted negotiations, it was abandoned.

2.   The Commission suggested a research fund to decrease manning requirements on ships and improve competitiveness.

3.   Technical standardisation of equipment for member states, in order to reduce conflict. This related to the overall Community theme of harmonisation.

4.   Social measures to improve working conditions, for example decreased hours, common training schemes and mutual recognition of qualifications. Again this related to harmonisation.

5.   Environmental action to ensure the same standards of marine pollution prevention in all ports.

6.   As an incentive to join EUROS, all vessels registered under the scheme would receive priority to transport surplus foodstuffs as Community aid.

7.   A Community shipowner was defined - an important issue to clarify as it affected cabotage rules.

8.   Removal of restrictions on cabotage, which related to liberalisation. Although, or perhaps because, it attempted to open up the market this was a highly politically contentious area. The Commission put forward the idea that all Community countries should open up their domestic shipping to EUROS vessels. Manning requirements would be the same as for the member state in question, which might conflict with EUROS requirements. It was proposed that this item be reviewed in January 1993. A "get-out" clause was provided in that each member state could define specific routes which required subsidy to operate and were needed for public service reasons. Subject to Commission approval, these routes could then be reserved to national carriers.

In December 1990 European Community Transport Ministers agreed that a first phase of liberalising cabotage was to begin during 1993, amid strong opposition from the Greek and Italian Governments and doubts over its legality. The ministers agreed that the first phase should cover

14

mainland cabotage and a second phase island cabotage.

Legal doubts surrounded the setting of a date "during 1993" - after the 1992 deadline - and the way that only "mainland" shipping would be included in the first phase of liberalisation.

In February 1991, lawyers confirmed that the agreement reached in December was in breach of the Treaty of Rome (Lloyds List 20/2/1991). However, the Commission's lawyers concluded that derogations from a January 1993 cabotage introduction could be made.

In December 1991 Lloyds List reported that European Community Transport ministers had agreed a framework for liberalising maritime cabotage and that they hoped to finalise a deal within the following six months. The new scheme envisaged liberalisation by January 1st 1993 of mainland cabotage in the tramp and liner sector, where a journey is seen as one leg of an international voyage. The second step would involve all other traffic between mainland ports. The last three stages cover island traffic and would begin with tramp and other liner traffic, before all remaining island traffic except passenger traffic, with the last stage being regular passenger ferries. Final agreement resulted in a cabotage deadline of 2004 for all trades in the EC.

Paper 2. "Financial and Fiscal Measures Concerning Shipping Operations with Ships Registered in the Community".
(Commission for the European Communities 1989).

Whereas the first document set out proposals for legislation, this second document was mainly advisory and concerned subsidy, its definition in a European context and the limitations acceptable to the EC.

The Commission recognised that Community shipping was already heavily subsidised and that this was distorting the market. However, the Commission agreed to approve certain limited subsidies to help to retain Community ships under Community flags and employ Community seamen. Subsidies were only to be allowed for social security payments, training and differential tax regimes.

On 15th April 1991 the Community drew up "Guidelines for the examination of state aid to Community shipping companies" (Commission for the European Communities 1991). These guidelines, which replaced the second document of Stage Two of the EC shipping policy, defined ten types of vessel which could receive different levels of subsidy. The aim of the plan was to restrict subsidies to their lowest levels.

Another paper outlining a new EC maritime policy was drawn up by Dr Bangemann, Vice President of the European Commission. According to Dr Bangemann the advent of the single market within the EC will double the present volumes of cross border transport - the vast majority of it by sea - by the year 2000. At the same time international trade, 90 % carried by shipping, was expected to continue growing (Lloyds List 29/8/91).

To prepare the EC for the upturn, Bangemann advocated a European Maritime Agency, primarily as a forum for promoting co-operation between all parts of the industry and EC governments. He dismissed previous maritime policy as "out of date" (Lloyds List 30/8/91).

In an interview with Lloyds List on 3rd September 1991 Dr Kroeger, Managing Director of the Association of German Shipowners, stated that the test for an effective European Maritime policy is "whether it offers conditions which would make the industry want to stay in Europe". Although commenting that Bangemann`s proposals were welcome he felt that they had not gone beyond a "very initial phase".

On September 28th 1991 it was reported in the Telegraph that the European Commission had set up a forum to examine Community maritime policies and to find ways of making EC shipping, shipbuilding and maritime services more competitive. The 'Maritime Industries Forum' was asked to produce a report and recommendations by summer 1992 (Telegraph 28/9/91) and since then has continued to meet periodically.

Since the recent economic and political changes that have taken place in East Europe, the maritime area of that region has received little direct attention from the EC. Whereas in the early stages of formulation of a maritime policy, the Community saw East European countries as a competitive threat (ESCEC 1977) this is no longer the situation. The main concern of the EC now is how to help former members of the CMEA survive the difficult transitional period. It is in the interests of the Community that pluralist democracy and a market economy should prevail throughout Europe (Pinder 1991), and the Community must consider what it can do to help East Europe towards this situation, and how in this process to aid East European shipping as part of economic redevelopment. In the meantime, other issues have taken a greater role, for example the Single European Market, harmonisation and liberalisation, and European Monetary Union, and these temporarily have relegated East European maritime issues.

New members

With recent changes taking place throughout Europe the actions of the EC look increasingly likely to be influenced by new members and this in turn will have clear impacts on the shipping sector. By 1996 the EC consisted of the fifteen Member States of Austria, Belgium, Denmark, Finland, France, Germany, Greece, Ireland, Italy, Luxembourg, Netherlands, Portugal, Spain, Sweden and UK, although this number looks set to rise.

A report in The Independent (1/5/1991) stated that the EC insisted that it would not consider any applications before 1993, making accessions before 1995

16

unlikely. Despite this there has been some frenetic activity in terms of applications.

On 4/4/1991, it was reported in the UK Independent newspaper that President Lech Walesa hoped preliminary negotiations would be able to admit Poland as an associate member by the end of the year and to full membership soon after. On December 16, 1991 Poland entered into an associate agreement with the EC. However, Community sources described the timetable as over optimistic and said although they hoped preliminary negotiations would have been tied up by December, there were several problems that would need to be dealt with before associate status was possible.

On 21 April 1991 the European newspaper reported that Malta hoped to join the EC. Meanwhile, the collapse of the Soviet Union had caused the Finnish economy to plunge. In 1985 the Soviets bought 21.5% of Finland's exports; in 1991, this was expected to fall to less than 5%. The worse Finland's recession became, the keener it became to join the EC (The Economist 12/10/90). The Finnish Prime Minister recognised that:

> The Europe we are in is changing, and changing dramatically and a sober assessment of our national interests demands that we are part of the EC (Financial Times 6/3/92).

A direct consequence of this was that the government applied for EC membership in March 1992 (Lloyds List 19/3/92), and acceded along with Austria and Sweden in 1995.

East European countries' hopes of joining the EC were heightened in May 1991 with a radical suggestion from within the EC executive that a form of flexible affiliation could solve many of the problems blocking membership. The idea seems to suggest that criteria for EC membership must be loosened if they are to be appropriate for developing economies (Lloyds List, May 1991).

Meanwhile a significant trade deal was signed on 22nd October 1991, with the intention of creating the world's largest 'barrier free' market covering 380 million consumers in 19 countries (Financial Times 23/10/91). After months of negotiation, the EC and the (then) seven members of the European Free Trade Association (EFTA) - Finland, Iceland, Norway, Sweden, Switzerland, Liechtenstein and Austria - finally agreed to the creation of a European Economic Area (EEA).

The new agreement was to bring EFTA into the Single Market from the end of 1992, with its unrestricted movement of goods, people, services and capital, although not agriculture, fisheries, energy, coal or steel.

In return EFTA nations were asked to provide financial aid, as well as to accept, uncritically, some 10,000 pages of EC regulations covering trade, competition and financial services.

Early beneficiaries of the EEA could have been Spain and Portugal, which were enjoying an economic boom based on low labour costs and which together with Greece and Ireland would receive substantial economic aid from EFTA in the package. But investments could quickly have switched to even poorer countries in Eastern Europe should the EEA expand eastward (Lloyds List 24/10/1991).

Despite continued problems relating to the legality of the agreement, by February 1992 the final details had been agreed. The success of the EEA negotiations raised hopes that the new democracies of East Europe would have a smoother path to membership (Lloyds List 28/10/1991).

The EC set up a working group to analyse the impact of the EEA on shipping. The group was to focus on the likely effect on coastal trades in the region. Also, it was decided that most EC maritime rules would be applied to maritime industries in the EFTA countries, though it was understood there would be certain anti trust alterations (Lloyds List 24/10/91).

Expanded membership of the EC however is highly probable despite the EEA development, although it is likely to cause a number of problems. For example agreements between such a large number of countries with such varying backgrounds will be difficult, and international relationships will become increasingly complex.

Negotiations concerning a common shipping policy for the EC are likely to be further hampered and the rate of progress may be slowed considerably as agreements become more difficult and time consuming to reach. The possibility of East European countries joining the EC would affect Community attitudes, and would alter the development of any potential shipping policy.

Single market

In order to become full EC members East European countries would need to accept the Treaty of Rome which calls for the creation of a market economy with minimal state intervention, limited restrictions on freedom to provide services, and no discrimination on the basis of nationality, i.e. the principles of harmonisation and liberalisation. The Treaty envisaged that the Community's prosperity and in turn its political and economic unity would depend on a single integrated market (Commission for the EC 1991).

Commitment to achieving a single market by the existing member states was reflected by the signing of the Single European Act (SEA), leading to the formation of the single market by 31st December 1992. The single market was defined by the act as being "an area without frontiers, in which the free movement of goods, persons, services and capital is assured in accordance with the Treaty of Rome". (EC Commission 1989). The SEA came into force on 1st July 1987, and introduced changes to the Treaty of Rome such as the abandonment of unanimity required to pass EC legislation. The requirement became qualified majority interest except where vital national interests prevailed. Since then more rapid progress has been made in achieving aims, and many original single market proposals have been approved by the Council.

The six objectives of the SEA were outlined by the Commission (1991) as follows:

1   The completion of a large market without frontiers, allowing the free movement of goods, people, services and capital.

2   Structural policies pursued at Community level providing opportunities for all regions, particularly those which are economically behind.

3   Co-operation on research and technology.

4   Monetary co-operation, or the introduction of the European Monetary system.

5   To apply the social dimension which forms part of the Single European Act.

6   To develop and implement a European environmental policy.

In 1988 the gains of forming a single market were calculated at 216bn ECUs, 5.3% EC GDP and 1.8m new jobs, estimated as the price of the existing barriers. The basic feeling was that "eliminating the 12 separate markets and creating one, would create a circle of growth without inflation, with greater competition, forcing prices lower, forcing innovation, creating even greater competition" (Commission for the European Communities 1987). A single European market could lead to economies of scale, job creation, improved productivity, improved profitability, healthier competition, mobility, stable prices and wider consumer choice.

The creation of a single market by 1st January 1993 relied upon the removal of present market barriers by all members and the passing of much legislation by both the EC and the member states. Shipping legislation amongst other sectors soon fell behind the necessary rate of action.

There were three types of broad barrier to the creation of a single market; attempts to remove these affected shipping as follows (Commission of EC 1991; GCBS 1990; Hayer 1990; Lloyds Shipping Economist, February 1990):

1. Physical barriers including border checks and customs points. Attempts to remove these in shipping have included:

*   The introduction of the Single Administrative Document to ease movement of goods.

*   Common integrated customs tariffs for goods moving within or into the EC.

*   Collection of intra-community trade statistics. Adopted 1st December 1988, this means that from 1993 all importers and exporters will provide

EC data, backed by a customs inspection service.

* Banalisation - goods moving within the EC would require inspection at internal customs points only once.
* National technical conformance checks, so that border controls on technical detail will not be needed. This only marginally affects shipping as IMO and ILO conventions control technical operations.
* VAT proposals - allowing deferred payment. At present immediate payment is required which wastes administrative time and effort and creates cash flow problems.

In relation to passengers a few factors are relevant to shipping:-

* Controls on the internal movement of EC citizens will be eliminated.
* Arms and drugs control, immigration and tax relief need to be rationalised although this is causing some problems.

Generally shipping will benefit from reduced customs procedures and a general simplification of rules and paperwork.

2. Technical barriers relating to the technical specification of vessels, taxation and legal frameworks. These are being removed in three ways:

* Since 1983 the Commission has been able to over-ride any national technical legislation acting as a barrier to trade.
* Attempts have been made to regulate banking and insurance industries.
* Work is being carried out to produce mutually recognised standards of qualification.

The idea behind these measures is that goods lawfully manufactured and marketed in a member state should be allowed free entry into another state. This is directly relevant to the transfer of ships and marine equipment between states and is indirectly relevant to shipping as it will encourage the movement of goods between member states and hence commonly by ship.

Technical barriers to persons will also be removed where possible. Freedom to work anywhere in the EC is one of the basic rights of the Treaty of Rome. In terms of shipping, the Commission are working on two specific proposals:-

(i) Mutual recognition of seafarers' qualifications.
(ii) Harmonisation of training and supplies on board ships.

The EC also hope to remove national barriers to the sale of "financial products", mainly banking and credit, insurance and securities. Liberalisation of insurance throughout the EC would benefit shipping through greater choice and lower prices.

Transport represents 7% of EC GDP, and is essential to liberalise if many opportunities of the SEM are not to be lost. Although there has been some progress towards a common transport policy, especially in road haulage and air, shipping has seen little development, which is particularly emphasised by the problems of introducing legislation relating to cabotage and EUROS.

The Treaty envisages free movement of capital within the EC. This will be relevant to the shipping industry which requires large amounts of financial flexibility, whilst the EC also aims to remove difficulties for businesses operating in more than one member state by harmonising company laws. Since the shipping industry is inherently international and commonly operates over national boundaries this must have some effect.

There have also been various proposals to harmonise taxation rules in the EC. Of particular note to shipping is the proposal that rules for determining taxable profits should be harmonised, and that the Commission intends to spread the cost of a depreciating asset over its full life. It is also proposed to extend roll over relief on capital gains to 4 years as long as money is reinvested in ships, buildings or aircraft.

Further proposals include free competition and state aid harmonisation. These issues are relevant to shipping and are tackled to some extent in Stages One and Two of the EC shipping policy.

3. Fiscal barriers to shipping relating to monetary rates and excise duties on moving goods internationally. Attempts to remove these barriers have included those relating to public procurement. It is estimated that 15% of EC GNP is spent by governments. The EC would like to see open tendering, so that the Government of any member state, could be supplied by nationals of any member state.

Article 99 of the Treaty of Rome recognises the need to adopt measures towards the harmonisation of legislation concerning turnover, taxes, excise duties and other indirect taxation to harmonise for the internal market. Direct effects on shipping include:

- VAT on passenger transport.
- Loss of duty free facilities.
- VAT applied to a wide range of ships purchased for the first time.

Although some progress has been made towards the creation of a truly single market since 1992, shipping is one of the areas in which the EC is well behind schedule. The amount of shipping legislation passed by the EC in order to create a common policy, has been limited, as have been its effects upon the countries of East Europe over which the EC has no jurisdiction. However, if East European countries were to become members of the EC they would need to adopt this EC legislation and apply the general liberating culture of the Single European Market.

Conclusions

Since the early days of the development of an EC shipping policy, when concern about East European shipping activities was evidenced by the reports of Prescott (1976-77), Couper (1977), ESCEC (1977) and Seefeld (1977 and 1979), it is noticeable that East Europe has declined in its significance as other issues have become dominant and the recent dramatic changes in East Europe, for example the accession of Gorbachev and the collapse of the Council for Mutual Economic Assistance (CMEA), have called for a change in the attitude of the EC towards East Europe generally, as well as specifically in relation to shipping issues.

Looking at the European maritime scene, an article in Lloyds Shipping Economist, February 1990, highlighted that "it is likely that the main issues will remain unresolved for a long time yet" (Anon 1990). This indicates that the situation regarding the relationship between EC and East Europe and their shipping policies is likely to remain unsettled for the time being. Although EC progress towards a common policy has been slow, Stage Three of an EC shipping policy seems likely to emerge eventually; the effect that this has on East Europe will depend upon whether East European countries decide to join and are accepted by the EC. According to Merritt (1991) the East Europeans' eventual membership of the EC is inevitable. During his EC Commission Presidency, Jacques Delors repeatedly emphasised the point to political leaders across Europe that if the rest of the Community wants Germany to remain firmly anchored inside the EC, Eastern Europe cannot be left outside it. The turning towards the West of the former Comecon countries marks a fundamental shift in relationships within Europe, which will undoubtedly have impacts upon the shipping industry.

Although East Europe has declined in its significance to the EC recently in terms of shipping policy, a number of major issues in this area will remain of interest to the EC. These may include the consideration of Western access to East European markets. Also there are still likely to be some concerns about East Europe's need to earn hard currency and the methods by which this is achieved; for example by undercutting western shipping operations. Linked to this is that there are concerns over hidden maritime subsidies in East Europe leading to anti-competitive practices. The EC may also be interested in the advance in technology and efficiency of Eastern shipping industries which may be achieved through the introduction of schemes such as privatisation or joint ventures.

These broad maritime issues are likely to remain of some interest and concern to the EC until a better understanding and a closer relationship develop with East Europe.

References

Arbuthnott H. and Edwards G. 1979 *Common Man's Guide to the Common Market.* Macmillan Press Ltd.

Bergstrand S. and Doganis R. 1987 *The Impact of Soviet Shipping.* Allen and Unwin.

Commission for the European Communities 1987. *Single European Act* (passed 1st July).

Commission for the European Communities 1989. *Paper 1. A future for the EC Shipping Industry: Measures to improve the operating conditions of Community Shipping.* COM (89) 266 FINAL 3 August.

Commission for the European Communities 1989. *Paper 2. Financial and Fiscal measures concerning shipping operations with ships registered in the Community.* SEC (89) 921 FINAL 3 August.

Commission for the European Community 1991 (June). *Opening up the internal Market.*

Commission for the European Communities 1991. *Guidelines for the examination of state aid to Community shipping companies.*

Couper A.D. 1977, Shipping policies of the EEC. *Maritime Policy and Management.* Vol 4, No.3, pp129-140.

Erdmenger J. and Stasinopoulos D. 1988 The Shipping policy of the European Community. *Journal of Transport and Economic Policy.* September, pp355-360.

ESCEC 1977 *EEC's transport problems with East European countries.*

European 21/4/91 Malta doesn't want to be an island.

Financial Times 6/3/92 Finns set to follow dream of the reluctant European.

GCBS 1990 Joint Government/ Industry Working Party on British Shipping: Submission by the GCBS on fiscal issues. HMSO.

Hayer J. 1989 Unpublished MSc Dissertation. Plymouth Polytechnic.

Independent 4/4/91 Walesa seeks early Polish entry to European Community.

Independent 1/5/91 Switzerland to aim for EC membership.

Lloyds List 16/1/92 EUROS bids for June start.

Lloyds List 20/2/91 EC cabotage plans face a rethink.

Lloyds List, May 1991 East Europeans may be 'affiliated' to EC.

Lloyds List, 22/10/91 New Swiss parliament confronted by EC issue.

Lloyds List 28/10/91 Europe's trade deal heartens EC applicants.

Lloyds List 19/3/92 Finland applies for European Community membership.

Lloyds List 29/8/91 Bangemann sees a new maritime era.

Lloyds List 30/8/91 Saving EC industry through growth and co-ordination.

Lloyds List 3/9/91 Effective European Maritime Policy Vital.

Lloyds List 24/10/91 EC to check impact of EEA on shipping.

Lloyds List 19/11/91 European register set for adoption.

Lloyds List 2/12/91 Progress on EC cabotage.

Lloyds List 18/12/91 EC agrees on cabotage plan.

Lloyds Shipping Economist, Feb 1990 EC shipping policy flagging.

Lloyds Shipping Economist 12/10/90 Between two worlds.

Merritt, G. 1991 *Eastern Europe and the USSR*. Kogan Page Ltd.

Pinder, J. 1991 *The European Community and Eastern Europe*. Pinter Publications.

Prescott J.L. 1976 Community Shipping Industry, EC European Parliament Working Documents.

Russell W.R. 1975 Press release on a speech by retiring chairman of the Council of European and Japanese National Shipowners (CENSA) in London on 11/12/75.

Seefeld, H. 1977 *Interim Report on Sea Transport Problems in the Community*. European Parliament Document 5/77.

Seefeld, H. 1979 *Report on the present state and progress of the common transport policy*. European Parliament Working Document 512/78.

Times 2/7/91 Sweden formally applies to join EC.

Treaty of Rome 1957.

Simulation of a shipping line with a ferry feeder service

Ludwik Kondratowicz
Institute of Maritime Transport and Seaborne Trade
University of Gdansk

Abstract

The paper presents a multimodal transport-oriented simulation modelling system of a simple example of a simulated transport network that involves one ocean shipping line, a short-sea link, road transport, and two seaports.

Main features of the simulation system

Simulation experiments were carried out by means of *MULTIMOD*, the Simulation Modelling System for Multimodal Transport Logistics - a decision-support expert system, which enables the modelling, simulation and optimization of any existing or planned seaport or an inland freight transport terminal, or a group of interconnected terminals, large or small, multi-purpose or specialized, at optional scale. The system makes it possible to simulate and optimize the functions of a specific transport mode (maritime, road, pipeline, etc.), or combined, intermodal transport including production, storage and distribution, according to the rules of logistics. Hence, *MULTIMOD* makes it possible to generate transport simulation models structured exactly for whatever the current purpose might be and according to specified assumptions.

The *MULTIMOD* system provides a powerful simulation technique while eliminating the costs, delays and risks of modelling and programming that typify simulation projects in transport applications. As distinct from traditional simulation modelling, which combines data, knowledge and control programming, the data-driven simulation modelling approach of the *MULTIMOD* system treats data and general control logic as distinct parts. Use of the system requires only an understanding of the seaport (or inland terminal), or the transport system under study. Different transport configurations can be simulated by changing the input data, either current or stored in library data files, as the knowledge base.

Benefits of the system include its ease of use, flexibility and user friendliness. The models are self-documentary; the interactive modelling process is rapid as the user gets results quickly with no programming delays. Automatic logical control of the inputted model, checking on its coherence and consistency according to inference rules of the domain knowledge are also performed.

An important feature of *MULTIMOD* is that a goal function can be freely defined and a series of simulation experiments (runs) are performed connected with automatic goal function evaluation, which speeds up analysis of the simulated system and helps to identify the optimal solution. Results of the simulation enable comprehensive and multidimensional analysis of either the whole intermodal transport system, or the port or inland terminal against a transport framework.

The *MULTIMOD* modelling philosophy treats seaports and terminals as key logistics centres of intermodal freight transport. It supports evolutionary modelling of transport systems and their environment. Either macro or micro analysis of the simuland can be carried out. Some portions of the system functions, or a flow of cargoes and vehicles, can be studied at a detailed level, and others at a coarser level. For these reasons *MULTIMOD* is particularly suitable for logistics analysis in strategic, tactical and operational decision-making.

The *MULTIMOD* system, being a multimodal transport-oriented tool, can be utilized in a wide area of applications, while the modelling process is fast and simple, and does not require special skills and computer programming.

MULTIMOD as an expert system, comprises inference mechanisms which support the decision-maker during the modelling phase and the simulation phase.

The use of *MULTIMOD* is reasonably simple as there is no mysterious language to learn and no programming to master. Without extensive training, one can execute models that will enhance decision-making ability and help lower operating expenses. Complete, detailed models of a transport or terminal system are built by describing the system components and the logic of its operation using a clear interactive, user-friendly question-reply procedure. *MULTIMOD* models are then simulated to see how they react to changes in design or operation, just as the actual logistical chain or its subsystem would react. *MULTIMOD* models can predict the impact of changes on the existing system or a proposed design. This results in an integrated system-wide view of how any change impacts upon system performance.

The simulation model and results of simulation

The subject for this simulation is the movement of cargoes in the Baltic Sea area where ocean shipping lines cooperate with a local feeder line. The simulation model, fully deterministic, is intentionally hypothetical and simplified because its main purpose is to demonstrate selected modelling techniques and approaches that are possible by means of the *MULTIMOD* system.

The following vehicles were included in the model:

1 *Seaqueen*: a deep-ocean liner vessel which arrives at the Port of Gdynia every 168 hours.
2 *FerryGH*: a ferry-boat operating between the Port of Helsinki and the Port of Gdynia.
3 *LandGhW*: land transport delivering cargoes from the hinterland of Gdynia to be exported to the West.
4 *LandWGh*; land transportation for moving cargoes imported from the West, from Gdynia to the hinterland of Gdynia.
5 *Land HhGW*: land transport delivering cargoes from the hinterland of Helsinki, to be exported to the West via Gdynia.
6 *LandWGHh*: land transport for moving cargoes imported from the West, from Helsinki to the hinterland of Helsinki.

The following cargoes are transported:

1 *HtGdyW*: containers from the hinterland of Gdynia for export to the West.
2 *WGdyHt*: containers imported from the West to be delivered to the hinterland of Gdynia.
3 *HelGdyW*: containers from Helsinki for export to the West *via* Gdynia.

4 *WGdyHel*: containers imported from the West to be delivered to Helsinki via Gdynia.

There are two storage places for import and export cargoes identified in the model: *areaGdy* in the Port of Gdynia and *areaHel* in the Port of Helsinki.
   Cargo-handling equipment is represented in the model in very aggregated form, i.e. by two berths: one named *berthGdy* situated in the Port of Gdynia and the other *berthHel* in the Port of Helsinki. The berth in Gdynia is shared by the ocean vessel *Seaqueen* and by the ferry-boat, *ferryGH*, although they cannot use it at the same time.
   This data input produced the following results:

COMPUTER PRINTOUT OF THE MODEL

Model elements are printed in bold and italics.

MULTIMOD System Release 5.03   2.7.1995   10:54

MODEL STRUCTURE

MODEL AND SIMULATION CONTROLS

Model name: *NORTH-BALTIC FERRY-BOAT FEEDER SERVICE VIA THE PORT OF GDYNIA*

Length of the simulated period: 720 hrs
No. of repeats of the simulation period: 1 x

CARGO-HANDLING EQUIPMENT

#	Equipment name	Availa bility	Cost/work time unit	Cost/idle time unit
1.	*berthGdy*	1.	400.00	300.00
2.	*berthHel*	1.	350.00	300.00

STORAGE PLACES AND LOCATIONS

#	Storage name	Capacity	Utilized storage cost	Empty storage cost
1.	*Gdynia*	0	0	0
2.	*areaGdy*	80000.	5.00	3.00
3.	*Helsinki*	0	0	0
4.	*areaHel*	70000.	6.00	3.00

CARGOES
  1. *HtGdyW*   2. *WGdyHt*   3. *HelGdyW*   4. *WGdyHel*

INITIAL VOLUMES OF CARGOES AT STORAGES

Storage name	Cargo name	Storage units per cargo unit	Init.stock in storage units	Initial stock in cargo units
areaGdy	*HtGdyW*	1.00	3000.	3000.
areaGdy	*WgdyHt*	1.00	3000.	3000.
areaGdy	*HelGdyW*	1.00	3000.	3000.
areaGdy	*WgdyHel*	1.00	3000.	3000.
areaHel	*HelGdyW*	1.00	2000.	2000.
areaHel	*WgdyHel*	1.00	2000.	2000.

VEHICLES AND OPERATIONS

1.   *Seaqueen* - external vehicle.
      Arrival freq. (hrs): fixed = 168.000
      First arrival at: .0 hrs.

Cost in terminal: 1500.00/1hr.

Op.1/1:unload *WgdyHt* in *Gdynia*.Vol.fixed = 800.000
Op.2/1:unload *WgdyHel* in *Gdynia*.Vol.fixed = 600.000
Op.3/2:load   *HtGdyW* in *Gdynia*.Vol.fixed = 900.000
Op.4/2:load   *HelGdyW* in *Gdynia*.Vol.fixed = 1100.000

2.   *FerryGH* - internal vehicle.
      First generation at: .0 hrs.
      Single *FerryGH* operates in the system
      Cost during trip: 1400.00/1hr.
      Cost in terminal(s): 1100.00/1hr.
      Maximum cargo capacity: 500.0

Op.1/1:unload *WGdyHel* in *Helsinki*. Vol. as loaded before
Op.2/2:load *HelGdyW* in *Helsinki*.Vol.limited by max. cap.
Op.3/3:trip from *Helsinki* to *Gdynia*
Op.4/4:unload *HelGdyW* in *Gdynia*. Vol. as loaded before
Op.5/5:load *WGdyHel* in *Gdynia*.Vol.limited by max. cap.
Op.6/6:trip from *Gdynia* to *Helsinki*

3.   *LandGhW* - external vehicle.
      Arrival freq. (hrs): fixed = 8.000
      First arrival (or generation) at: .0 hrs.
      Cost during trip: .00/1hr.
      Cost in terminal(s): 300.00/1 hr.

Op.1/1:unload *HtGdyW* in *Gdynia*. Vol. fixed = 20.000

4.   **LandWGh** - external vehicle.
     Arrival freq. (hrs): fixed = 8.000
     First arrival (or generation) in: 4.0 hrs.
     Cost during trip: .00/1 hr.
     Cost in terminal(s): 350.00/1 hr.

Op.1/1:load **WgdyHt** in **Gdynia**. Vol. fixed = 25.000

5.   **LandHhGW** - external vehicle.
     Arrival freq. (hrs): fixed = 8.000
     First arrival (or generation) in: .0 hrs.
     Cost during trip: .00/1hr.
     Cost in terminal(s): 360.00/1 hr.

Op.1/1:unload **HelGdyW** in **Helsinki**. Vol. fixed = 30.000

6.   **LandWGHh** - external vehicle.
     Arrival freq. (hrs): fixed = 8.000
     First arrival (or generation) at: 5.0 hrs.
     Cost during trip: .00/1hr.
     Cost in terminal(s): 330.00/1 hr.

Op.1/1:load **WgdyHel** in **Helsinki**.Vol.fixed = 20.000

TRANSPORT SERVICE SYSTEMS

1.   **SeaqunGd**: **WGdyHt** from **Seaqueen** in **Gdynia** to **areaGdy**
     Cargo-handling speed:fixed = 50.000 units/1 hr
     Equipment required: **berthGdy** 1.0

2.   **SeaqunHe**:**WGdyHel** from **Seaqueen** in **Gdynia** to **areaGdy**
     Cargo-handling speed:fixed = 50.000 units/1 hr
     Equipment required: **berthGdy** 1.0

3.   **SeaqldGd**:**HtGdyW** from **areaGdy** to **Seaqueen** in **Gdynia**
     Cargo-handling speed:fixed = 60.000 units/1 hr
     Equipment required: **berthGdy** 1.0

4.   **SeaqldHe**:**HelGdyW** from **areaGdy** to **Seaqueen** in **Gdynia**
     Cargo-handling speed:fixed = 60.000 units/1 hr
     Equipment required: **berthGdy** 1.0

5.   **FerunWGH**:**WGdyHel** from **FerryGH** in **Helsinki** to **areaHel**
     Cargo-handling speed:fixed = 60.000 units/1 hr
     Equipment required: **berthHel** 1.0

6.   **FerldHGW**:**HelGdyW** from **areaHel** to **FerryGH** in **Helsinki**
     Cargo-handling speed:fixed = 55.000 units/1 hr
     Equipment required: **berthHel** 1.0

7. **FertrHG**:trip of **FerryGH** from **Helsinki** to **Gdynia**
   Trip time:fixed = 40.000 hrs

8. **FerunHGW**:**HelGdyW** from **FerryGH** in **Gdynia** to **areaGdy**
   Cargo-handling speed:fixed = 60.000 units/1 hr
   Equipment required: **berthGdy** 1.0

9. **FerldWGH**:**WGdyHel** from **areaGdy** to **FerryGH** in **Gdynia**
   Cargo-handling speed: fixed = 60.000 units/1 hr
   Equipment required: **berthGdy** 1.0

10. **FertrGH**:trip of **FerryGH** from **Gdynia** to **Helsinki**
    Trip time:fixed = 35.000 hrs

11. **LanunGhW**:**HtGdyW** from **LandGhW** in **Gdynia** to **areaGdy**
    Cargo-handling speed: fixed = 20.000 units/1 hr

12. **LandldWG**:**WGdyHt** from **areaGdy** to **LandWGh** in **Gdynia**
    Cargo-handling speed:fixed = 20.000 units/1 hr

13. **LanunHGW**:**HelGdyW** from **LandHhGW** in **Helsinki** to **areaHel**
    Cargo-handling speed:fixed = 18.000 units/1 hr

14. **LanldWGH**:**WGdyHel** from **areaHel** to **LandWGHh** in **Helsinki**
    Cargo-handling speed: fixed = 20.000 units/1 hr

END OF MODEL

COMPUTER PRINTOUT OF THE RESULTS OF THE SIMULATION RUN

MULTIMOD System (Release 5.3)   2.7.1995   10:55

RESULTS OF THE SIMULATION RUN #   1

NORTH-BALTIC FERRY-BOAT FEEDER SERVICE VIA THE PORT OF GDYNIA

Model file: feeder.mod. Observations file: feeder.obs.
Simulation of 720.0 hrs period completed (repeated 1.0 x)

REPORT ON CARGO-HANDLING EQUIPMENT

Equipment/ restrict. name	No.of times used	Cumul. work time	Util. (%)	Total cargo	Unfulfilled demand serviced (# times)
berthGdy	28	365	50.7	20500	544
berthHel	9	79	10.9	4500	0

31

USE OF STORAGES

Storage name	Cargo name	Delivered	Taken	Min. stock	Max. stock	Mean stock	Utiliz (%)
areaGdyHt	**GdyW**	1820	4500	260	3080	1698.5	2.12
	WgdyHt	4000	2250	2950	4850	3911.6	4.89
	HelGdyW	2000	5000	0	3000	756.3	.95
	WgdyHel	3000	2000	:000	4000	3696.1	4.62
Total:		10820	13750				12.58
areaHelHel	**GdyW**	2730	2500	1450	2340	1900.5	2.71
	WgdyHel	2000	1800	1700	2440	2057.4	2.94
Total:		4730	4300				5.65

MOVEMENT OF VEHICLES IN TERMINALS

Vehicle name	Arrived No.of rotations	Serviced queue	Mean queue	Max queue	Final	Cum. waiting time of not serviced veh's
Seaqueen	5	4	.0	0	0	.0
FerryGH	1	4	.3	1	1	45.6
LandGhW	91	90	.0	0	0	.0
LandWGh	90	90	.0	0	0	.0
LandHhGW	91	90	.0	0	0	.0
LandWGHh	90	90	.0	0	0	.0

TERMINAL TIME ANALYSIS OF VEHICLES WHICH LEFT TERMINAL(S)

Vehicle name	Cum. time in terminal	Mean time waiting	Longest time waiting	Cumulat. time waiting	Mean time in terminal	Longest time in terminal
Seaqueen	245	61.3	61.3	0	.0	.0
FerryGH	334	37.2	75.6	189	21.0	58.9
LandGhW	90	1.0	1.0	0	.0	.0
LandWGh	113	1.3	1.3	0	.0	.0
LandHhGW	150	1.7	1.7	0	.0	.0
LandWGHh	90	1.0	1.0	0	.0	.0

```
TRANSSHIPMENTS REPORT SORTED BY CARGOES
- - - - - - - - - - - - - - - - - - - - - - - - - - - - - - - - - - - - - - - -
Cargo        Vehicle        Delivered            Taken
name         name
- - - - - - - - - - - - - - - - - - - - - - - - - - - - - - - - - - - - - - - -
HtGdyW       Seaqueen           0                 4500
             LandGhW          1820                   0
                              --------           -------
Total:                        1820                 4500


WGdyHt       Seaqueen         4000                    0
             LandWGh            0                  2250
                              --------           -------
Total:                        4000                 2250




HelGdyW      Seaqueen           0                 5000
             FerryGH          2000                 2500
             LandHhGW         2730                    0
                              -----              -----
   Total:                     4730                 7500



WgdyHel Seaqueen              3000                    0
        FerryGH               2000                 2000
        LandWGHh                0                  1800
                              ------             ------
   Total:                     5000                 3800
```

TRANSSHIPMENTS REPORT SORTED BY VEHICLES

Vehicle name	Cargo name	Delivered	Taken
Seaqueen	*HtGdyW*	0	4500
	WGdyHt	4000	0
	HelGdyW	0	5000
	WGdyHel	3000	0
		------	------
Total:		7000	9500
FerryGH	*HelGdyW*	2000	2500
	WgdyHel	2000	2000
		------	-------
Total:		4000	4500
LandGhW	*HtGdyW*	1820	0
LandWGh	*WgdyHt*	0	2250
LandHhGW	*HelGdyW*	2730	0
LandWGHh	*WgdyHel*	0	1800

TRANSPORT SERVICE SYSTEMS PERFORMANCE

System #name	No.of times active	Cumulat. process time	Total cargo served	Cargo served/ time	No. of times system idle due to lack of cargo/storage/equip		
1 *SeaqunGd*	5	80	4000	5.56	0	0	0
2 *SeaqunHe*	5	60	3000	4.17	0	0	106
3 *SeaqldGd*	5	75	4500	6.25	0	0	0
4 *SeaqldHe*	5	83	5000	6.94	9	0	110
5 *FerunWGH*	4	33	2000	2.78	0	0	0
6 *FerldHGW*	5	45	2500	3.47	0	0	0
7 *FertrHG*	5	200	2500	3.47	0	0	0
8 *FerunHGW*	4	33	2000	2.78	0	0	328
9 *FerldWGH*	4	33	2000	2.78	0	0	0
10*FertrGH*	4	140	2000	2.78	0	0	0
11*LanunGhW*	91	91	1820	2.53	0	0	0
12*LandldWG*	90	113	2250	3.13	0	0	0
13*LanunHGW*	91	152	2730	3.79	0	0	0
14*LanldWGH*	90	90	1800	2.50	0	0	0

					9	0	544
%	1.6				0		98.4

COST ANALYSIS OF PROCESSES

System name	Cumulated process cost	Cost per work time unit	Cost per cargo unit
SeaqunGd	32000.0	400.0	8.0
SeaqunHe	24000.0	400.0	8.0
SeaqldGd	30000.0	400.0	6.7
SeaqldHe	33333.3	400.0	6.7
FerunWGH	11666.7	350.0	5.8
FerldHGW	15909.1	350.0	6.4
FertrHG	280000.0	1400.0	112.0
FerunHGW	13333.3	400.0	6.7
FerldWGH	13333.3	400.0	6.7
FertrGH	196000.0	1400.0	98.0

Total:	649575.8		

COST ANALYSIS OF EQUIPMENT

Equipment name	Work time costs (ewt)	Idle time costs (eit)	Equipment cost (ec) (ewt+eit)	ec:tec %	tec:gtc %
berthGdy	146000.0	106500.0	252500.0	53.45	
berthHel	27575.8	192363.6	219939.4	46.55	
	----------	---------	---------		
Total: (tec)	173575.8	298863.6	472439.4		24.84

COST ANALYSIS OF STORAGE PLACES

Storage name	Utilized storage cost(sut)	Empty storage cost(sem)	Storage cost(sc) (sut+sem)	sc:tsc %	tsc:gtc %
areaGdy	452.8	8368.3	8821.1	50.17	
areaHel	244.3	8517.9	8762.1	49.83	
	--------	--------	--------		
Total: (tsc)	697.1	16886.2	17583.3		.92

```
COST ANALYSIS OF VEHICLES
-----------------------------------------------------------------
Vehicle      Trip costs    Terminal      Vehicles      vc:tvc  tvc:gtc
name         (vtr)         costs         cost (vc)       %        %
                           (vte)         (vtr+vte)
-----------------------------------------------------------------
Seaqueen           .0      368000.0      368000.0      26.06
FerryGH      476000.0      418000.0      894000.0      63.31
LandGhW            .0       27000.0       27000.0       1.91
LandWGh            .0       39375.0       39375.0       2.79
LandHhGW           .0       54000.1       54000.1       3.82
LandWGHh           .0       29700.0       29700.0       2.10
             ---------     ---------     ---------     -------
Total:       476000.0      936075.1     1412075.0     100.00
(tvc)
```

End of simulation results

No attempt is made here to interpret the results of this analysis - merely to present them as an example of an application of the model.

Other possible applications of the *MULTIMOD* system

MULTIMOD can be used to simulate functions of numerous systems and applications. Here are just a few examples:

- Generate and evaluate various alternative logistical chain scenarios.
- Seaports and inland terminals, both multipurpose and specialized.
- Influence of different transport technologies, technical developments, modal splits, or seaports and terminals.
- Shipping lines with feeder networks.
- Analysis of the competitive position of short sea shipping against road transport.
- Bulk cargo shipping operations.
- Inland waterway transportation.
- Road transport network functioning.
- Detailed container terminal operations.
- Evaluation of future changes in the direction and intensity of traffic corridors.
- Production - storage - transport complex, which may include, e.g. a factory, refinery, intermodal transport, inland depots, storage tanks, and seaports - all as an interrelated whole.
- Parallel simulation of two or more seaports or inland terminals, which are mutually dependent, e.g. sharing common hinterlands, resources, vehicles,

servicing policy.

- System integration between parties involved in sea-land transportation.
- Verification of investment projects in seaports and shipping.
- Evaluation of broken chain concepts versus unbroken chains, intermodal transport versus direct.
- Urban transportation systems and interchange terminals.
- Warehousing and distribution systems.
- Repositioning of containers in the intermodal transport network for the purpose of maximization of their utilization.
- Changing availability of the resources over time, for various reasons.
- Restrictions on the simulated system functioning due to labour rules and different work organization systems (shifts, length of weekend).
- Restrictions on terminal operations because of randomly changing weather conditions, tides, channel traffic and locks.
- Random or scheduled arrival patterns of all types of vessels, trucks, trains, barges, or other vehicles freely defined by the modeller. Seasonal variations. Optional individual control of single vehicles movements beyond the terminal(s).
- Cargo-handling equipment breakdowns and maintenance.
- Alternative transport technologies.
- Various stocking systems, alternative handling technologies for all types of cargoes.
- Complex servicing priorities of various vehicles, cargoes, allocation of resources and alternative transhipment techniques.

References

Kondratowicz L.: *Computer simulation in logistical analysis of multimodal transport. Methodology and implementation.* University of Turku, The Center for Maritime Studies, Turku, Finland, 1994.

Polish seaports in the service of ferry shipping in the Baltic sea region

Konrad Misztal
Institute of Maritime Transport and Seaborne Trade
University of Gdansk

Abstract

This paper examines the development of Polish ferry ports in the Baltic Sea region and in particular the specific facilities available in each, and the new proposals put forward for their improvement, concentrating upon the provision of new road facilities.

The high concentration of short sea ferry and ro ro lines is a characteristic feature of the European market. About 330 ferry connections are in service around Europe. The North Sea and the Baltic Sea have the highest density of ferry shipping among the European Seas. The Polish ferry fleet operates on the Baltic Sea, although its share in Baltic ferry shipping is rather small - it amounts to ca. 1% in passenger transportation and ca. 5% in cargo transportation.

The Baltic ferry market comprises transportation of passengers, rolling and rail cargo, feeder services and cruises for passengers. At the same time, that market is a segment of the European transit market in a North-South direction. It embraces the countries of Scandinavia, Central-East-West Europe and the Near East. That market constitutes a transport network linking Scandinavia with West and Central Europe - a network which is used for transporting the majority of cargo and passengers in the region.

The central location of Poland in Europe, especially its location on the European North-South route, along with comparatively small distances from the Scandinavian countries are the features which predispose Polish ports to serve as the base ports for ferry and ro ro lines connecting Polish seaports with Scandinavia. Many routes have already been redeveloped e.g. Swinoujscie-Ystad and Swinoujscie-Copenhagen ferry lines, Gdansk-Oxelosund ferry lines, Gdynia-Helsinki-Kotka ro ro line, Gdynia-Karlskrona and Gdynia-Stockholm connections served by Swedish shipowners. To service the above mentioned lines, Polish seaports operate ferry and ro ro terminals in Swinoujscie, Gdansk and Gdynia.

The first ferry terminal was opened in Swinoujscie in 1964. In the beginning it was served by one passenger-car ferry sailing on the Swinoujscie-Ystad line, then in 1978 the second rail-car ferry was put into operation. Reconstruction and modernisation of that terminal was undertaken in 1988, and both are expected to be finished in 1996. Besides modernisation of two existing berths, three new ferry berths are under construction. In 1995 a new car terminal, with 11 departure places and parking, will be put into operation thus ensuring proper services for both passenger cars and trucks. The terminal is able to handle 4-5000 passengers and 15 passenger-car ferries daily. It also possesses the only wharf for rail-car ferries in Poland. The terminal can serve 88,000 passengers, 150,000 passenger cars, 100,000 trucks and 60,000 carriages annually.

In Gdansk the main ferry terminal is situated on the Ziolkowski Quay, 300m from the entrance to the port. The length of this quay is 132m. Parking can accommodate at the same time 130 cars, 12 road sets and 19 containers of 40 ft. The terminal possesses only one berth for handling a passenger-car ferry of 3,000 RT. Nearby, on the Oliwski Quay there is a one stand ferry terminal for passenger-car ferries of a gross capacity 12,000 RT, with a handling capacity amounting to 0.4 million tonnes yearly. This berth can also handle ro-ro, con-ro and lo-lo vessels.

Summary of container ro ro, lo lo and ferry cargo capacity of Polish seaports.

Gdansk

	Container lo-lo and ro-ro	Ferry ro-ro
No. of cargo ship visits	221	
Tons of cargo handled	1426t	
Numbers in TEUs	115	
No. of berths available	1	1
Berth spaces available	900m	130m
Storage space	20,000sqm	
Equipment	2 multi purpose Kone cranes 2 shore gantry cranes of 40t	full back up facilities
Max. draft	9.6m	6.5m

Gdynia

	Container terminal	Ferry terminal
Daily through put capacity	1,200 TEUs	20,000 passengers an hour
Berth space available	798m	178m
Max. draft	10.1m	8.1m
Sheds	25,000sqm	
Open storage yards	63,300sqm	
Equipment	3 35/40t capacity gantry cranes, 14 stacker cranes, rubber-tyred gantries, 3 straddle carriers, 1 multi-purpose crane, various tractors	
Container handling facilities	2 adjustable ramps for ro-ro vessels, the container freight station, modern cold store, bonded storage for cars, containers and paper	

Szczecin

There are five quays with suitable equipment for handling passenger vessels and cargo traffic, including warehouses of 1,500sqm and bonded storage:

	Ferry Terminal
Quay No. 6	Capable of handling cargo vessels (ro-ro and lo-lo) length 130m, max. draft 7m
Quay No. 5	Capable of handling passenger ferries, ro-ro vessels and cruisers; equipped with a ramp for cars and a gallery for boarding passengers, length 200m, max. draft 7m
Quay No. 4	Capable of handling passenger ferries, ro-ro vessels and cruisers; equipped with a ramp for cars and a gallery for boarding passengers, length 206m, max draft 5.6m
Quay No. 3	Capable of handling car/train and passenger ferries, ro-ro vessels and cruisers; equipped with 2 loading ramps for tracks and wagons, length 197m, max. draft 5.6m
Quay No. 2	Capable of handling car/train and passenger ferries, ro-ro vessels and cruisers; equipped with 2 loading ramps for tracks and wagons, length 183m, max. draft 7m

Source: Polish Ports Handbook 1995. Maritime Economy and Industry Guide. Szczecin.

The port in Gdynia has at its disposal two ro-ro berths in the container terminal. One of them has been adapted (with its immediate land hinterland) for handling passenger-car ferries of 150m length and 5.4m draught. The second berth serves a ferry line linking the Stockholm agglomeration - operated by a Swedish ship owner 3 times a week. In the port in Gdynia there are also favourable conditions for handling ro ro and con ro vessels in the Baltic Container Terminal. The ferry terminal in Gdynia is of a provisional character - it was constructed as a temporary solution for handling ferry traffic in this port. Its location within the container terminal not only limits the territorial development of the base but also does not provide communication connections free from the likeliehood of collisions (especially the road connections with transport infrastructure of the city and hinterland.

Changes which have taken place in the Baltic Sea region create new possibilities for operation and development of Polish seaports. However, the specific location of Polish ports has unquestionable ascendancy over their place and position in the Baltic Sea region. They are located rather peripherally, outside the main oceanic shipping routes and outside the main industrial centres of Europe. Hence, temporal and economic distances are unfavourable for many cargoes. As a consequence, Polish ports will undoubtedly never gain a position of logistic and distribution centres on a European scale (and maybe not even on a regional scale) as the base ports for European shipping and centres for cargo concentration.

A steady concentration of oceanic trade in several base ports (the so called 'sea ports polarisation') made the Baltic ports serve as regional and feeder ports, supplying cargo for the base ports. The base ports of the Baltic ocean trade are large West-European ports such as Hamburg, Bremen, Rotterdam and Antwerp. In turn, feeder services are being developed on the Baltic Sea. The only port which has kept its oceanic position is the Swedish port of Gothenburg. The other Baltic ports have become regional ones, adapting themselves to the new role by constructing specialised ro ro berths, terminals for feeder services and regular Baltic lines.

In the light of this tendency, Polish ports may look for development possibilities in re-opening Baltic shipping connections with Swedish, Finnish and Dutch ports, along with opening new shipping lines to the ports of Latvia, Lithuania, and Estonia. One chance of attracting cargo and passengers to Polish ports may derive from development of the European North-South transport corridors, running from Scandinavia, via the Gdansk and Szczecin agglomerations, to Southern Europe and further to the Near East. The favourable location of Poland in that system may be conducive to the development of sea-land transit on the North-South route using sea routes linking Polish ports with Scandinavian countries.

With a view to the development of the Polish transport corridor, great importance should be attached to the POLLINK* system embracing the new Baltic chain - a ferry connection between the ports of Stockholm and the Gdansk agglomeration. POLLINK is the shortest ferry-rail and ferry-road connection for passengers and cargo between Finland, South-East Sweden, via the ports of Gdansk and Gdynia to Southern Europe. The Trans European Railway (TER) and Trans European Motorway (TEM) constitute the land chains of the corridor. Within the TER and TEM projects, it is planned to construct modern railway and road connections running from Scandinavia, via the seaports of Szczecin and Gdansk agglomerations, to the Central/East European countries, and to the Mediterranean countries. Development of these corridors will influence cargo and passenger movements in the Scandinavia - South Europe - Near /East directions - and for this direction Poland is considered to be a natural transit route. It will

*The sea-land North-South transport systems comprise two other systems: SCANLINK and DANLINK. The SCANLINK system will constitute one large transport system linking Scandinavia with the continent. Its basic elements will embrace rail and road connections between Oslo, Gothenburg, Copenhagen and Hamburg - these links will be established due to the opening of regular ferry connections in the Dutch Straits. In turn, DANLINK is a joint venture of Swedish, Dutch and German railways - tending towards the improvement of transportation passing in meridian directions.

result in the growth of both multimodal transport (ro ro and ferry shipping) and international tourism, thus influencing the development of passenger movement. Changes which occurred in shipping in the Baltic Sea region force Polish seaports to adapt themselves to the new situation. A foreseen growth in ferry transportation and in ro ro turnover in that region mean that Polish ports face new tasks with respect to quality and quantity. Hence, besides the modernisation of the biggest ferry terminal in Swinoujscie, it is planned to construct a regular ferry terminal in Gdynia and to build a new ferry terminal in Gdansk (in the Eastern area which is to be linked with the Trans European North-South Motorway). New terminals for handling ro ro turnover are planned to be constructed in Gdansk, Gdynia, and Szczecin.

The place and position of Polish seaports in the Baltic Sea region are determined not only by their location in relation to the main shipping routes and industrial centres of Europe, but also by the changes that are taking place in the European transportation systems. In relation to that, the position of Polish seaports seems to consist of:

1 Performing the function of feeder ports for containerised general cargo - which requires development and construction of ro ro berths and port terminals for serving feeder lines to the base ports in North West Europe, along with opening new connections with Baltic ports.
2 Performing the function of transit ports in the system of European North-South corridors running from Scandinavia via Polish seaports to Southern Europe and further to the Near East - thus taking over cargo and passenger movements in that direction.
3 Performing the function of transhipment ports for the Baltic countries (Lithuania, Latvia and Estonia) whose transhipment capacity is not sufficient for handling containerised, ro ro, lo lo and other specialised cargo.

References

Ferry Shipping on the Baltic. Budownictwo Okretowe i Gospodarka Morska. No. 11, 1991.

Polish Ports Handbook 1995. Maritime Economy and Industry Guide. Szczecin.

System of transport connections between Polish ports and Central European countries with a view to establishing the Adriatic-Baltic Hexagonal Group and possibilities of its utilisation in transit services. Maritime Institute. Gdansk.

Wisniewski E. Polish Ferry Shipping - Characteristics and Development Premises. Budownictwo Okretowe i Gospodarka Morska, no. 6, 1991.

Analysis of
Polish short sea
shipping

Zofia Sawiczewska
Institute of Maritime Transport and Seaborne Trade
University of Gdansk

Abstract

This paper takes a long and detailed look at the role and importance of short sea shipping based in Poland in terms of the Polish economy and its significance in the international market place, providing recommendations for the future.

The rapid changes which have occurred during recent years in Eastern and Central Europe have resulted in an entirely new political and economic situation. Increasing trade between Poland and Scandinavian countries necessitates the development of an efficient short sea shipping and inland transport corridor for cargo and passenger traffic. If the current situation in Poland is analysed, and the forecasts for 1996 by leading international institutes are taken into account, there are some reasons for optimism.

In 1995 and 1996 real economic growth of 4.5% is expected, although - in terms of unemployment growth - this scarcely relieves the labour market and only contributes slightly to solving the social problems. Taking into account other forecasts for Polish foreign trade, an annual growth rate of 3-4% to the year 1995 can be anticipated.

Considering all the macro-economic data there is reason to hope that within the decade, countries such as Poland, the Czech Republic, Slovakia and Hungary will be close to gaining the status of countries of the European Union.

Multimodal and logistic management approaches need new ideas and concepts based upon prospects for the future. Altogether there is a considerable potential for development of seaborne trade and passenger traffic within the Baltic region as well as in the external traffic in and from the destinations outside the Baltic Sea.

In the sea transport area the most important factor arising from the Baltic countries is the increase in efficiency of ferry and ro-ro connections in the central and eastern transport corridor of the Baltic. In Poland supportive infrastructure is the major obstacle to attempts to achieve more efficient short sea shipping services - as cargo dispersion is often delayed due to missing links from ports to the internal transportation system.

Lack of capacity stems mostly from an inadequate domestic transportation system and problems of interface between the domestic and international sectors.

Several points concerning short sea shipping and the national intermodal system need to be made clear to the decision making bodies:

- The system must be much more than a map, network, or inventory of facilities.
- It must include safety, efficiency, environmental aspects and physical conditions of the system components.
- Identification of the components requires a basic elementary approach together with both the local and state transport decision making.
- The system must focus on connectivity between modes and on intermodal facilities.
- The system must preserve the ability of the freight transportation sector to operate its privately owned infrastructure efficiently.
- The system must recognise the need for fundamental change in the government's control and financing institutions.
- One must stress again that the importance and adequacy of intermodal

connections are the weakest points. They are, generally speaking, the links in the current transport system and the points of transfer between individual transport means.

- Because the present system is financed and managed separately, responsibility for strengthening these links is unclear but necessary.
- Also the relationship between the system and computer information technology must be emphasised.
- The opportunity to use information systems for network capacity management, telecommuting and information exchange that reduce congestion by substituting for transportation, must be considered.

The era of intermodal freight transportation began in the mid 1980s when ocean carriers and railroads teamed up to launch double stack rail container services. This approach stacks two shipping containers on specialised railcars for greater efficiency. Since the system was introduced in the USA and around the world growth has been explosive.

New partnerships between ocean carriers, railroads, truckers and shippers have been formed and efficient, cost effective services have brought dramatic changes to land and ocean shipping. The new intermodal partnerships among rail, truck and ocean carriers offers lower costs which result in lower prices for consumers, improved marketability for exporters and congestion relief as well as environmental benefits, because rail and sea transport cause less damage to the environment.

Poland can be considered late in these development and therefore the administration system needs to be reconstructed to achieve the intermodal vision for the future. The present modal structure of the government and local administration is a barrier to development of a national intermodal transportation system together integrating with the European short sea shipping concept.

The administration has traditionally served as advocate for specific modes and as a result intermodal and European global concepts and projects have been neglected. In addition, co-operation with the government was frequently inadequate, as policy was not directed to benefit the maritime sector in Poland.

Shipowners also have had difficulties in determining their policy for the situation in Europe after 1992 and in Poland after 1989. Rules and regulations were unknown and shipowners were too deeply involved in changing their ownership and internal structures. Poland's insufficiently developed financial system and burdensome taxes have also been slowing the maritime sector's development that should power the country's growth.

It seems that rather than helping or even ignoring shipping enterprises, government policies were, in practice, starting to get in their way. Obviously without restructuring in companies - which is actually taking place - and reduction of government constraints upon the business environment, Poland's maritime industry will not be able to continue playing a dynamic role in the country's drive to catch up economically with industrialised countries of the West.

Shipping companies cannot still rely on retained earnings to fund growth and cannot hope that their shares go public. In addition, increasing competition has squeezed margins so thin that covering costs has become difficult for the business. Obviously it is no longer enough to say 'the government should not disturb the industry, it can simply develop on its own'. If the Polish government wants the maritime sector to continue driving the economy, policy makers need to create more hospitable conditions, and policy recommendations need to be formulated to promote the European short sea shipping concept in Poland.

Polish foreign trade statistics indicate that in 1993 Polish exports amounted to 13,878million US$, whereas imports reached a level of 15,878million US$. The trade deficit for 1989 was US$ 2,293million. Meanwhile, the relative importance of the EU in Polish trade has increased significantly. On the exports side the share increased from 28% in 1989 to 58% in 1992. Germany has become the most important trade partner with its share at 31% of Polish exports in 1992. Meanwhile, on the import side the comparable shares of the EU were 31% and 53%. The largest share of Polish imports in 1992 again originated from Germany - 24%.

The most substantial share of Polish seaborne trade traditionally went to western European countries (about 30%) and to the Baltic region (about 20%), the remaining 50% allocated to deep sea traffic.

About 17% of the total cargo turnover was transit cargo coming to and from the former Czechoslovakia, Hungary and Austria through Polish territory. According to recent Polish shipping statistics, in 1993 the total amount of cargo carried by the Polish fleet reached 23,869,000 tonnes. The share of western European seaborne trade reached 887,000 tonnes in 1993 for regular shipping connections compared with a total of 7,088,000 tonnes.

The share of Western European seaborne trade in the bulk trade was 8.59m tonnes compared with the total of 24.5m tonnes In 1993 the Baltic carriages were 1.74m tonnes for regular shipping and 0.367m tonnes for the bulk trade.

Passengers are an important cargo in short sea traffic with the following carriages reported:

In 1992 - 674,853 passengers were carried by ferries, 4,685 passengers were carried by other ships.

In 1993 - 626,166 passengers were carried by ferries, 3,582 were carried by other ships.

On the Sweden - Poland route 445,000 passengers were carried by ferries in 1991, 547,000 in 1992 and 517,000 in 1993.

On the Finland - Poland route 41,000 passengers were carried in 1991, 29,000 in 1992 and 26,000 in 1993.

On the Denmark - Poland route 94,000 passengers were carried in 1991, 81,000 in 1992 and 95,000 in 1993.

The following loadings have been reported by the main Polish shipping companies:

Polish Ocean Lines - 2,157 passengers in 1990, 1,713 in 1992, 489 in 1993.
Polish Steamship Company - 2,157 passengers in 1990, 1,437 in 1992, 1,393 in 1993.
Polish Baltic Company - 560,452 passengers in 1990, 538,806 in 1992, 491,752 in 1993.
Euroafrica - 2,798 in 1993.
Corona Line - 133,121 in 1993 (suspended in 1994).
POL Levant - 153 in 1993.
POL America - 51 in 1993.

All shipping companies in Poland are involved in short sea shipping, but only some operate ferries and/or tramp ships on the short ranges exclusively, that is around the European continent as well as on the Baltic and the North Sea. These are:

Polish Baltic Shipping Company with 16 ships (21,358t) in 1990, 18 ships (24,655t) in 1991 and 18 ships (25,464t) in 1992. According to statistics on the 31st December 1992 the company was operating 7 ferries (9,183t) and 11 small bulk carriers (16,277t) exclusively in short sea ranges.

Euroafrica, the first independent shipping company in Poland (with 50% of shares in private hands) established on 1 September 1991, operating to the UK, Scandinavian and Western European countries, but also reaching out to the West African Coast. At the end of 1992 Euroafrica achieved carryings of 1,322,000 tonnes in short range services and 264,000 tonnes to West Africa.

Next is Pol-Levant Shipping Lines, which has operated since May 1993 on the Mediterranean route around the European continent as a private company with Euroafrica having the majority of its shares.

Also, Polish Ocean Lines has been involved in short sea shipping, offering ro-ro services to the Scandinavian countries and the UK. One could observe a slight tendency to increase the share of short sea carriages in POL's services and there is some possibility of increasing their role in the future.

Meanwhile, there were 11 ferries (71,754BRT, 20,719t) in the Polish fleet at the end of 1992 and 10 ferries (68,739 BRT, 18,684t) in 1993.
The Polish Steamship Company in Szczecin (PZM), which specialises in the bulk trades, has also been involved in short sea shipping, having at the end of 1992, 225 (353,720ton) bulk carriers, and 12 conventional tramp ships (49,025ton) exclusively engaged in the European ranges.
Looking for reasons for development of ferry services one must mention the trade structure of the Baltic Sea on one hand, and the special transport conditions

which prevail on the other. Whereas external Baltic transport is mainly raw materials, internal trade is dominated by passengers and finished and half finished goods.

This is one of the reasons why traditional Baltic shipping has been replaced by ferry and ro-ro services. The demand for the quality of services is constantly increasing as the value of the goods increases and more and more industries are switching over to a logistical approach and just in time production methods.

Meanwhile, state activities in Poland in transport are being extended to cover superstructure. The erection and maintenance of the infrastructure is still regarded as being the responsibility of the government. In terms of infrastructure, the best known, long term project of the modern railways network (TER), requiring sizeable investment, aims to establish high standard railway connections for both passengers and cargo services. This centres around multimodal links, implementing European consignment and wagon tracking together with a full information network available to all passengers.

Meanwhile, in terms of the contribution of inland waterways, two large Polish rivers, the Wisla and Odra, theoretically the cheapest and cleanest way of transportation across the country, serve only about 1% of the cargo to Gdansk and about 3% to Szczecin, due to their poor navigation infrastructure.

At present, in the Baltic region, the TEM/Trans-European North South Motorway project is underway, together with the TEM Scandinavia development. Starting in Finland, it would connect ferry links to Estonia, Latvia, Lithuania, and to the TEM in Poland.

The Via Baltica is a North South transport corridor through the Baltic countries, connecting to Finland in the North, Poland in the South, and further on into Central Europe. The route serves traffic in and between the Baltic countries, and provides a new link between North-Eastern and Central Europe.

The route extending from Tallinn to Warsaw is nearly 1,000km long. At present the lowest volumes (2,000 vehicles per day) are at the international borders and the highest (over 10,000 vehicles per day) in the vicinity of the bigger cities along the route. Immediate actions, such as border crossing imrovements, maintenance and minor road improvements will cost a total US$40million for the entire route over two years. Larger investments in road and bridge construction are proposed at a total of 105-230 million US$ for the five year programme.

For both projects (TER and TEM), credits and support from the World Bank have already been obtained. Another motorway, the Via Hanseatica would follow the Southern Baltic coast line up to Germany. It will connect the old Hansa towns and the developing transport and trade centres of the Southern shores of the Baltic Sea: Kiel, Hamburg, Lubeck, Rostock, Szczecin, Gdansk, Elblag, Kalingrad, Siauliai , Riga, Tartu, Tallinn, Narva, and St Petersburg.

Additional expansion is required to connect Polish motorways with Byelorussia, Western Ukraine and the Baltic republics. In addition to the currently existing roads, 1,500km needs to be modernised or constructed. Work has begun on priority areas, including road sections at the border crossings at Kalingrad - Elblag (Russia-Poland) and the Narva by-pass (Estonia-Russia). The construction of the via Hanseatica will speed up the transport of goods and passengers in the East-West Baltic-Hansa transport corridor. This project will become one of the major areas of international economic cooperation in the Baltic Sea region in the years 1995-2010. It will also be one of the most promising investments in the development of industry, construction, transport, trade and tourism.

Another project - The 'Kwiatowski Road' - is being developed to provide direct access from the Baltic Container Terminal in Gdynia to the main road connecting Gdynia with Gdansk and with Szczecin through the Gdansk ring-motorway. This road, essentially the beginning of the projected Trans European Motorway North-South, is a vital link within the entire road network, stretching from Scandinavia via Central Europe to Turkey. One of its accesses would connect Gdynia Eastern Port to the main motorway system helping to upgrade the port related services, and in the process, enhancing the prospects for Gdynia's economic development.

A part of the Kwiaktowski Road project is being financed by the World Bank. The required technical documents have already been completed; the binding procedure started in April 1995 and construction began in September 1995.From the environmental point of view construction is potentially favourable. Completion of the road would support general port development, especially with respect to the construction of a large passenger terminal and the attraction and settlement of industries.

Some of these concepts go together with the concept of short sea shipping development, some may appear as competition. However, European short sea shipping has an opportunity, if competitiveness, in the form of ro ro cargo handling, continues to improve, together with full mechanisation of loading and unloading processes. The dynamic development of Polish cargo traffic, far exceeding the present capacities of inland road infrastructure shows the necessity for such future developments. Each day a large number of private and newly privatised truck companies fiercely compete on the Polish transport market, in spite of the underdeveloped road infrastructure and ecological dangers. All these must be persuaded to consider the alternative presented by the European short sea shipping concept.

However, if most investment is allocated for the development of road transport, this reduces the resources available for the stimulation of short sea shipping. Also, the Polish ports, particularly Gdansk and Szczecin, are unfriendly for short sea trade, being essentially geared to deep sea trade. Generally, all three large Polish ports are functioning well and the real bottleneck was not the ports themselves, but the inadequacy of the inland transport system. At the same time other smaller harbours have been totally forgotten or connected with fisheries only. They could surely play much more important roles in the future, where short sea shipping is concerned.

In this respect there is a need for better market information for both shipowners and ports, to be able to make better market predictions as well as for legal support, brokerage facilities and financial management in the future to encourage the growth of short sea shipping trade.

Policy recommendations

Sea transport along European coasts can constitute an interesting alternative for an ecologically dangerous and too densely used road system. Ferry traffic in particular, can play an important role in the Baltic region and within other European ports. A coastal and short sea shipping system should be a system consisting of a combination of road, rail and sea transport.

A fundamental principle for the system must be the view of shipping as the main agent of a transportation chain which will take goods all the way to the receiver. With the shipping technology available today, in the form of vessels, handling systems and modern systems of organisation, shipping can increase its competitiveness against inland transport only marginally. It is necessary to decrease the cost for the transfer of goods from shore to the ship, to introduce a regular 24 hour service and to minimise time in port.

The in order to achieve this will require a completely mechanised cargo handling system. Only a system with mechanised loading and unloading, based on the use of the ro ro mode would, in the future, be cost and time competitive with inland transport modes. Also, the focus should be on improvements in packaging, storage, transhipment and a need for an effective intermodal management of the logistic chain. The system should also provide a good service level as regards numbers of departures, transport time, delivery reliability etc..

As part of this process, projects to assess infrastructural comparisons for Polish shipping costs need to be carried out and the costs of the short sea and inland (road) transport should be investigated and compared.

A new era of transportation is emerging in Poland, an era of "intermodalism", which refers to interconnections among modes of transport, use of multiple modes for a single trip, and coordinated transportation policy and decision making. It is vital to recommend ways to speed national conversion to an efficient intermodal transportation system and identify and allocate resources necessary to do it.

The government should consider all modes of passenger and freight transportation and focus on:
- The status of the existing intermodal system.
- Legal, regulatory and institutional issues.
- Funding and financial questions.
- Technology and research issues.

The views and expertise of transportation experts in the public and private sectors, as well as the general public, are needed. Necessary steps must be taken to ensure an intermodal transportation system that meets the needs of passenger and freight transport in the future, while recognising funding limitations and environmental considerations. The intermodal approach has much to recommend it, including more efficient use of the nation's transport infrastructure, better service, more convenience and more choice for users.

In time of fiscal difficulties, intermodal transportation brings many opportunities to gain maximum benefits from minimum resources spent. Transportation policy traditionally focused on single elements; automobiles, trains, trucks, ships, aeroplanes and transit systems. In an intermodal transportation system, these elements are connected in a system that is efficient, safe, flexible, environmentally friendly and meets the needs of the nation's travellers and shippers. Intermodalism offers the promise of:

- Lowering overall transportation costs by allowing each mode to be used for the portion of the trip to which it is best suited.
- Increasing economic productivity and efficiency, thereby enhancing global competitiveness.
- Reducing congestion and the burden of over stressed infrastructure components.
- Generating higher returns from public and private infrastructure investment.
- Reducing energy consumption and contributing to improving environmental conditions.

Various barriers exist to the development of a fully intermodal national transport system. Planning and policies do not encourage and accommodate either intermodalism or European short sea shipping projects.

In addition, funding of transport programmes is directed modally, discouraging investment in global intermodal transport and most government institutions are organised along modal lines, which inhibits planning and developing an intermodal transport system. On the other hand, not all intermodal transport and short sea shipping problems require government solutions. State policy should, and does, support private sector innovation, provide maximum flexibility for state and local transportation officials and not intrude unnecessarily into private sector operations.

There is a need for statistical data about transport flows between European regions. It is necessary to find out where volumes of flows are sufficient for investments in freight logistic systems, and also where a better integration with inland transport is required, ie. to make choices on the future long distance freight networks.

At present, short sea shipping is in an unfavourable position as compared to road transport. However, if some necessary improvements of the port and inland infrastructure are introduced, there will still be a chance to reduce the present focus on roads in favour of a more environmentally friendly traffic mode such as short sea shipping.

Real competition between sea and inland transport in Poland today is very limited. For goods in inland transport, the share exposed to competition is only approximately 1% of the total quantity of inland goods. This means that any increase of this competition will bring additional goods to shipping and at the same time the infrastructure already existing for shipping will be used to a greater extent.

The new transport system should be designed according to the conditions attractive to sea-land transport flows and based on sea-land transport units. The form of intermodal, wheel-borne cargo carriers should be developed with as great as possible number of options for on and out going transport connections within the European continent. This system should be based on the door to door services, offering at the same time the possibility of simple, intermediary storage of goods in ports.

One can predict, that at the very least, ferry traffic will become more important in the future, together with the development of road and rail connections. One can already observe the increasing mobility of people caused by income and free time growth. Tourism forms a major portion of such traffic, while business travel forms the rest. An important element of the attraction of ferry traffic is the opportunity for duty free shopping.

Another opportunity, which plays a growing role in northern and western Europe, is the development of ferry feeder services. Since 1980 the share of feeder traffic as part of total container traffic in northwest Europe has increased from 14.7% to 16.1% in 1990 and is expected to grow in the future. In absolute terms, feeder traffic is expected to grow to 3 million TEU in 1995 and to 4.1 million TEU in 2000 which means an average annual growth rate of 6.4%. Compared to this - the growth of the share of direct trades is forecast to reach 5.6% per year. Three major sources will contribute to the future potential of sea feeder traffic volumes performed by Polish shipping, and based on Bremen, Hamburg and London deep sea connections which already exist::

-   Normal seaborne trade development.
-   Generation of the new feeder traffic, especially between Gdansk, Gdynia, and the north European range.
-   Possible substitution of inland feeder flows between transit countries and the north European ports.

The following government and local municipal administration actions may succeed to support and encourage the development of short sea shipping in Poland:

-   Helping to collect necessary statistical data.
-   Harmonising transit regulations.
-   Reducing taxes on trades involved in the European short sea concept, together with recognising mutual customs seals and introducing common customs documents.
-   Working out and introducing a joint border, and custom control procedures, on the borders between Estonia, Latvia, and Lithuania.
-   Introducing a common system of transit control and exchange of information.
-   Supporting the shipowner's efforts to modernise and restructure short sea

fleets.
- Supporting initiation of research programmes in the field of marine transport system development in European short sea shipping.
- Supporting the restructuring programmes of seaports towards more efficient servicing of ships of limited sizes, e.g. developing small harbours along the Polish coastline.
- Initiation and support of cooperation contacts between short sea shipping and deep sea shipping.
- Supporting research in the field of better integration of short sea shipping into the door to door transport chains.

Whenever possible, favourable conditions under which short sea shipping can develop, should be supported and encouraged.

References

Barfuss K.M. Report on the Current Status of the Debate over European Integration and Eastern Expansion of the European Union. Paper presented at the UBC Conference. Gdansk 27-28 April 1995.

Gospodarka Morska. Przeglad statystyczny. 1994 Instytut Morski, Gdansk 1995.

Hume J. M. and Pinto B. 1993 *Prejudice and Facts in Poland's Industrial Transformation Finance and Development,* June.

Process of Changes in Ownership in Maritime Transport in Poland and East Germany. Gdansk University 1994.

Sawiczewska Z. Reconstructing Polish Ports and Shipping. *Maritime Policy and Management.* 1992 Vol. 19, No. 1 pp 69-76.

Sawiczewska Z. 1993 in *Transport and Economic Development in the New Central and Eastern Europe.* Bellhaven Press, London.

Adaptation of Gdansk agglomeration ports to the service of fast ferry connections

Stanislaw Szwankowski
Institute of Maritime Transport and Seaborne Trade
University of Gdansk

Abstract

This paper looks at a particularly interesting and topical new development in the region, in the form of fast ferries and their physical requirements in the seaports of the region. It reveals that the introduction of new technical facilities of this sort is likely in the market environment of the Baltic region and that the Polish ports will have a major part to play.

The Baltic Sea is characterised by a high level of activity of ferry and ro ro shipping, although density of the ferry connections is different for different regions of that aquen. Up to now, the central Baltic with the ports of the Gdansk agglomeration have been an area of relatively low intensity ferry passenger and cargo traffic. Economic, social, and political changes in central and west Europe and the aspirations of countries for integration within the Baltic sub-region are the features which create new development possibilities for ferry shipping linking Scandinavian countries with central/east Europe (via the ports of Gdansk and Gdynia). The process of Baltic European integration is likely to be followed by a serious growth of competition between ferry lines and from alternative land transport routes. The competition forces a new, wider outlook on the operation of a ferry corridor in the Gdansk and Gdynia ports. The development plans for those two ports should include the possibility of serving a new generation of high speed ships. The creation of handling possibilities for this type of modern tonnage in Gdansk and Gdynia will influence not only the improvement of the ports' competitiveness on the Baltic ferry market, but also their incorporation into services outside the Baltic and fast shipping connections to the ports of the North Sea and even to the English Channel.

Premises for serving ferry traffic in the ports of Gdansk and Gdynia

The ports of Gdansk and Gdynia constitute, beside Gothenburg in Sweden, the biggest ports complex on the Baltic Sea, of fundamental value to ferry shipping. They are centrally sited, not only within the Polish transport infrastructure, but also within the transport infrastructure of central/east Europe. Through Gdansk and Gdynia pass the shortest road and rail connections leading from Scandinavia to the main Polish urban centres - Warsaw, Lodz, Katowice, Cracow and to the economic centres of the central and east European countries.

In terms of the development of Baltic ferry shipping, the important advantages of Gdansk and Gdynia ports consist of:

1   Location advantages of the Gdansk agglomeration in the central part of the southern Baltic Coast, on the axis of the east European north-south transport corridor - which facilitates attraction of cargoes and passengers not only from central and eastern Poland but also from the potential hinterland of central and east Europe.

2   Favourable location of the two ports in the central junction of the planned TAPP motorway and fast railway connection (TER) with south Europe.

3   A large Gdansk agglomeration (over 1.5m inhabitants) with the necessary economic and intellectual capacity, as well as cultural and tourist/sightseeing

advantages of the whole Gdansk region (Kashubia regional park, Hel Peninsula, Vistula Bay).

About 80% of the whole country's hinterland is directed towards the Gdansk agglomeration ports - that area embraces the majority of industrial centres and approximately 80% of inhabitants. This hinterland consists of two segments of international transit hinterland:

1   A southern transit hinterland embracing Slovakia, eastern Czech Republic, Hungary and possibly the Vienna agglomeration of 27 million inhabitants; construction of the TEM will enlarge the territory of that hinterland by adding to it other countries sited along the TEM; i.e. the territory of the former Yugoslavia, Greece, Bulgaria and Turkey (with a population of 96 million).

2   Eastern transit hinterland which comprises Byelorussia, Western Ukraine and possibly even the Kalingrad region with a population of 30 million.

The possible centre for ferry connections with PAG (centring upon a range of operations of ferry connections with Swinoujscie) includes:

In the Baltic region

1   Ferry ports in central and eastern Sweden with two regions: the Stockholm agglomeration and Kalmar/Karlskrona, with the possibility of a connection with Gothenburg and Oslo by the E-77 route.
2   Ferry ports in Finland (Helsinki and Turku).
3   Ferry ports in the Baltic Republics of Lithuania, Latvia, Estonia, along with the Baltic ferry ports of Russia.

Outside the Baltic

1   In terms of high speed ferry connections the ports of the North Sea may also be considered.

Location requirements for ferry and ro ro berths in ports

Particular features of the ro ro reloading system determine the following local requirements, which must be taken into account when planning ferry terminals and berths:

-   Providing a possible short sea route (free from the risk of collisions) for ferries - leading from the port's entrance to the handling berth, especially in the case of short distance ferry lines.

- Providing land connections free from the risk of collisions (especially road) - which should be linked with the communication infrastructure of the hinterland and with the central areas of port's cities.
- Providing an adequately large hinterland enabling local parking and manipulation areas (ca. 2.0-2.5 hectare for 1 ferry berth).

The high speed vessel creates more requirements e.g. the requirement for minimum total layover time of the ferry in port influences the choice of terminal location (possibly near the port's gate and free from the risk of collisions within the port areas), its equipment and organisation of work.

With a view to the port's ferry terminals which handle passenger movement, their location is moreover conditioned by special requirements. When handling passenger movements a wide range of services are required: food, hotels, commerce etc, which tend to dictate location of ferry terminals in the central parts of ports' cities. On the other hand the generation of large amounts of traffic which arise from a ferry terminal, encourages the siting of ferry berths far from the city centres. This could lead to the segregation of passenger traffic into domestic and transit, and further may lead to the specialisation of passenger ferry terminals in the ports of the Gdansk agglomeration. At present, with a limited passenger movement the possibility of such segregation is rather small but as passenger transportation increases, these possibilities could become more and more realistic and hence they may constitute serious considerations when siting ferry berths.

Services for all kind of passenger movements have gravitated to the urbanised areas, these being visually attractive and well connected with the hinterland. Development of a port's capacity used for handling passenger movements is an action which stimulates territorial and economic integration between the port and city. Besides the advantages arising from present tourist traffic, maritime sightseeing and other projects could be developed around the character of the city, and enable passengers and citizens to see the port's activities and to observe the coast. Location of ferry terminals in the Gdansk agglomeration ports will be intrinsically tied in with the development of the centres of Gdansk and Gdynia - in terms of functional, organisational and territorial issues. This situation has occurred due to a strong 'town creative' character and strong influences on the development of the road network in the immediate and more distant hinterland.

Location of ferry terminals in the ports of Gdansk and Gdynia

The ports of the Gdansk agglomeration have at their disposal a rather limited technical capacity adapted to handling passenger car ferries. It includes a one stand terminal in Gdansk on the Capt. Ziolkowski Quay and the temporary ferry berth on the Helskie II Quay in Gdynia.

The ferry terminal in Gdansk was constructed in 1973. It is sited on the junction of two quays: Ziolowski and Oliwskie, 300m from the port's gate. Such a situation is a major advantage of the ferry terminal. The ferry berth has 125m length and 9m depth which enables handling of ferries of 4,000 BRT capacity.

Since 1989 the port in Gdynia has had at its disposal one ferry berth on the Helskie II Quay in the Baltic Container Terminal. The berth is adapted to handling passenger-car ferries of 153m length, 2m width and 5.4m draught. The berth has 190m length, a rather limited hinterland for passenger cars, and a passenger station of 800sqm. At present it serves ferries sailing on the Gdynia-Kariskrona route.

The new political and economic situation in the Baltic region encouraged the ports of Gdansk and Gdynia to prepare to construct new ferry terminals and to increase their activity in the Baltic ferry market place and particularly examine the potential of the new fast ferry mode.

The decision as to where the new ferry berths in both ports should be located resulted from an analysis of land use and the separate investments will be carried out in cooperation with both ports' cities and the ferry operator.

In the Gdynia port, several possibilities for ro ro and ferry berth sites had been considered over the last 20 years, usually under the influence of successive impacts which expected to arise due to the planned recovery of ferry shipping in Polish ports. In that time a new network of port locations had been established - including a container terminal, under construction on the Helskie Quay, which is adapted to container reloading using either lo lo or ro ro techniques. Recently that terminal was equipped with a new temporary ferry berth.

A major advantage of this scheme exists in the new communication network which is being developed. This network will pass outside the city centre and it will link the Helskie Quay territory with the Three-Towns peripheral motorway and with the wider Polish hinterland. The E. Kwiatkowski route is still under construction and although further investments are still necessary, it will create a chance to activate all port functions in the whole territorial system of that area (Basin VIII).

Construction of the multi-berth ferry terminal in the VIII Basin has the aim of serving passenger, car and rail movements. Investment is already planned by the Commercial Seaport of Gdynia SA, which is directly connected with the completion of the E. Kwiatkowski route (with the financial support of the World Bank) and its link with the peripheral motorway of the Gdansk Agglomeration.

Gdynia has other possible sites for ferry terminals. The configuration of the Dunskie Quay enables location of two ferry berths for handling car ferries or - in an alternative framework - designing one berth for rail-car ferry or ro ro vessels. The large subsidiary activities of the Dunskie Quay require provision of areas for storage, parking and handling. A favourable factor of this location is the possibility of linking the ferry berth with the existing Maritime Station by the E. Kwiatkowski route (under the stipulation that the problems of access and parking around the Maritime Station would be solved). The Dunskie Quay also has a favourable location with regard to the town centre.

In the port of Gdynia there are also other location possibilities for the ferry berth: on the Kutrowe Cuter and Wilson Quays - with their subsidiary industries in the central part of the Pomerian Quay. The first location, in the southern part of the fishery port, arises from the possibility of utilising these areas which could be redesigned in the course of the reconstruction programme. This region is favourably located with regard to the city centre, hence it should be developed either for the city creative function (e.g. logistic-distribution ones - eventually in connection with a Custom Free Zone) and for serving the passenger-tourist traffic (including ferry traffic) and for administrative/commercial and representative functions of the ports. In particular, location in this port's region is attractive for the port town - hence one could foresee that city authorities will press for placement of the tourist or commercial functions in these areas.

Similar advantages are attributed to the location of the ferry berth on the Pomeranian Quay. It is sited in the centre of Gdynia which is very attractive with respect to commercial and tourist functions. Moreover, it could be an ideal starting point for sightseeing tours for the whole territory of the Gdansk Agglomeration.

The functional-territorial system for the Gdansk port is based upon the location of separate port functions along the port channel - from which many basins deviate. Hence, siting of the ferry terminal is possible in different port areas.

The most favourable conditions for constructing a new ferry terminal in a short time are on the Oliwskie IV Quay. The stretch of 350m length provides possibilities of siting two berths for passenger-car ferries, each of 160m length. These sites are characterised by the proximity of the port's entrance and a modern circulation facility. Other possibilities for ferry berths in the Gdansk port have also been considered - in the Grain Quay region vis-a-vis the monumental Wisloujscie Citadel or inside the port in the Polish Hak region. The second arrangement is favourably sited in the centre of the historic town, but is far from the sea entrance to the port.

From the discussions above, it appears that the Gdansk and Gdynia ports have a number of location possibilities for modern ferry terminals. The potential schemes are characterised by benefits in respect either to their site with regard to the sea and local city area, and to the road connections with the immediate and more distant hinterland (free from the risk of collisions). The capacity of both ports exceeds predicted ferry traffic which is expected to reach 1 million tonnes of cargo and over 500,000 passengers in the year 2000. The volume of ferry transportation through the ports of the Gdansk Agglomeration will depend not only on the development of ferry capacity in both ports, but also on the marketing activity of Polish and foreign ferry operators, their competitiveness and on constructing a complementary road infrastructure.

References

Szwankowski S. and Tubielewicz A. 1994 *Investigation and marketing analysis of ferries' hinterland and forefield of the Gdansk Agglomerationports. Institute of Ecodevelopment Problems.* Ecobaltic Foundation. Gdansk.

Localisation programme conception of adapting the Gdynia port to handling increased ferry and ro ro traffic. The Maritime Institute. Gdansk.1991.

Land use study of the port of Gdansk carried out under the supervision of Tubielewicz A. The Maritime Institute, Gdansk. 1994.

The Polish passenger ferry industry in the post communist era

Neal Toy and Michael Roe
Centre for International Shipping and Transport
University of Plymouth

Abstract

This paper provides a study to identify and account for developments in industry during a period of substantial political, economic and social change in east Europe and aims to illustrate the major effects that these changes have had upon a particular sector of Poland's shipping industry; the passenger ferry sector. The broad approach is to outline the structure, conduct and performance of the ferry industry during communist rule; repeat this for the current situation and finally, account for the variations observed, if any are present. Many of the transitions relate to political, legal and structural entities and as such, by their very nature are more suitable for qualitative rather than quantitative appraisal.

To examine the industry in depth required the opinions and experiences of people within the industry, and a visit to Poland to carry out interviews was made in September 1994. Analysis and conclusions are based upon the use of free response interviews with industry and academic authorities, secondary quantitative data and published literature.

Introduction

The advent of Glasnost, Perestroika and the subsequent collapse of the Soviet Bloc, must figure as one of the most significant periods of the 20th century. In this early post-communist era, it has become apparent that many costs and benefits have been incurred, as countries struggle to discard one regime in favour of a fundamentally different one. Poland was one of the most eager to embrace the rapid transition towards political democracy and a market led economy. The adoption of a radical macroeconomic restructuring programme has caused short term hardships and political instability. At the same time the programme has ensured that Poland is relatively advanced in its transition when compared to other former Soviet Bloc countries.

This paper aims to illustrate the major effects that the economic and political changes have had upon a particular sector of Poland's shipping industry; the passenger ferry sector, with particular reference to the long established state operator Polska Zegluga Baltycka (PZB), who have experienced the transition from old to new. The broad approach is to outline the structure, conduct and performance of the ferry industry during communist rule; repeat this for the current situation and finally, account for the variations observed, if any are present.

Methodology

The objective of the paper, is to identify those areas of change within a shipping sector, which most significantly reflect the dramatic processes of political and economic transition within Poland since 1989. Many of the transitions relate to political, legal and structural entities and as such, by their very nature are more suitable for qualitative rather than quantitative appraisal, hence a general research approach was chosen (Patton 1987).

Although much work has been published on the social, political and economic transitions in Poland, and even on the effects on the shipping industry as a whole, the ferry industry has received only scant coverage. To examine the industry in depth, required the opinions and experiences of people within the industry, and a visit to Poland to carry out interviews was made in September 1994. Analysis and conclusions are based upon the use of unstructured interviews with industry and academic authorities, secondary quantitative data and published literature. In light of the dramatic and somewhat unique changes experienced by Poland, Professor J. Zurek of the University of Gdansk agreed to the general methodological approach undertaken, stating during interview: What is happening now is impossible to model or account for statistically. Statistics can only to be used for basic descriptive purposes (Zurek* 1994).

Professor Z. Sawiczewska supported this view (Sawiczewska*1994), stating:

> The economic and political change in Eastern Europe, the collapse of the Soviet Union together with German unification brought various questions which no contemporary economist can answer and situations which no country has experienced before (Sawiczewska, 1993).

Panel selection

The intention of the research process was to interview management from the state-owned company (PZB), existing and prospective new entrant companies and academics, in order to achieve a balance of viewpoints. Whilst the latter two groups were helpful, a lack of cooperation by PZB led to data and qualitative information for the company being derived from published material, trade journals, interviews with ex-employees and academic authorities. Although not ideal, this could not be avoided. An attempt at minimising obvious bias was made by the form and sequence of questions asked to others. Visits included Euroafrica SL, who provide the freight section of a joint service with PZB on the busy Swinoujcsie-Ystad route and, as such are in a position to provide a substantial input of information regarding the passenger operations of PZB both on that route, and in general. Also, through their involvement in the creation of an entirely new ferry company, Unity Lines, Euroafrica are soon to enter the passenger market themselves. Corona Lines were a new, privately owned Swedish/Polish joint owned company which operated from 1991/2 but which subsequently ceased trading.

The Polish shipping industry - during and after communism

Due to its turbulent political history, Poland has a relatively brief maritime history. It wasn't until 1918 that it gained the use of a section of its present Baltic Sea coastline. Post war restructuring of territories expanded the coastline and port infrastructure. The merchant fleet was immediately placed under state control and in 1951 the whole shipping sector was nationalised and organised to conform to the operating principles of a centrally planned economy (Ernst and Young 1990). Profits from shipping enterprises were absorbed by a centralised government body for redistribution according to the central investment plan. Shipping concerns therefore, did not always benefit from the income that they had generated, and consequently, investment in new tonnage slowed, as funds were redirected to higher priority areas.

Additionally, central government sought to delineate the shipping industry, which led, during the 1950s and 60s, to the predominance of two major shipping companies, each with well defined areas of operation, namely:- Polish Ocean Lines (liner) and Polska Zegluga Morska (tramping in the bulk sector). The process was finalised on January 1st 1970, when the government officially decreed the reorganisation of the shipping industry, whereby organisations were assigned exclusively to a single sphere of operation. This intensely monopolistic structure reflected the general "command" ethics of the Polish government towards all sectors of industry. In 1976 a new state owned operator, Polska Zegluga Baltycka (Polish Baltic Steamship Co), based at Kolobrzeg was instituted. Its principal role was to operate ferry services thus allowing Polish Ocean Lines to release this responsibility and concentrate on the liner market.

The role of shipping within the command economy

Fallenbuchl states:

> The expansion of the Polish fleet has always been regarded as a device designed to help the balance of payments position of the country, especially with transactions with the West. The expansion of the domestic fleet was regarded not only as an import substitution measure - which would reduce payments for foreign transportation services in connection with the Polish foreign trade and the passenger traffic between foreign countries and Poland - but also as an earner of foreign currencies (Fallenbuchl 1980).

Fallenbuchl however, does not account here, for the historical sequence of change in emphasis of the role of the shipping sector within Poland's post-war economy.

During the 1960s the emphasis began to change towards the independent role of the shipping services on the international market, as optimum fleet utilisation was not possible with purely CMEA trade and many wasteful ballast journeys resulted (Chrzanowski 1977). This factor was compounded during the late 1970s by Poland's acute shortage of foreign exchange brought about during the Gierek era. By the end of the decade, the ability of a shipping company to maximise foreign exchange was used as the principal measure of performance (Zurek* 1994). As shipping companies became increasingly autonomous during the 1980s, profit maximisation became established as the principal indicator of company performance (Ernst and Young, 1990).

Industry structure

As noted earlier, shipping in Poland was represented in recent years by three large state-owned companies with a specialised range of activities:

Polish Ocean Lines (POL), based in Gdynia, operated liner vessels since 1970.
Polska Zegluga Morska (PZM), based in Szczecin, a bulk tramp operator with worldwide operations.
Polska Zegluga Baltyka (PZB),largely a ferry operatorin the Baltic market.

These principal enterprises are still almost exclusively under state control and it is thought that they are likely to remain so for some considerable time (Zurek 1994).
There has been recent legislation to relax the delineation of the industry and allow the formation and entrance of entirely new enterprises into shipping markets. The Law on Economic Activity of December 23rd 1988, amended 1990:

..grants the right to enter into business activities to every body on the principle of equality, subject only to those restrictions which are specifically stipulated in the law (Ernst and Young 1990).

These liberalisation measures have already begun to de-centralise the industry structure.

Privatisation of the shipping sector

Privatisation was central to Balcerowicz's macroeconomic restructuring programme towards a western style market economy, and it is the key element of reform (Ledger and Roe 1993). Sawiczewska clearly illustrates that there were however, serious practical difficulties confronting all sectors of the Polish economy, not only the shipping industry, in the implementation of privatisation:

The post-communist Polish Government wants to create a Western structure of ownership and to return more than half of the state's assets into private hands. It appears a task of great complexity, as after 40 years of centralised manipulation, it is no longer clear who owns what, what the value of anything is, how it can be sold and to whom. There is a shortage of capital, expertise, proper management, motivation and efficiency (Sawiczewska 1992).

Ledger and Roe (1993) summarise the main incentives of Poland's privatisation as being; the need to encourage industry efficiency through exposure to competition; the need to reduce the size of the public sector, and; to encourage the spread of assets widely amongst the population. Additionally, revenues generated through the sale of state assets could be used to ease substantial foreign debts. Although the shipping industry was more familiar with western procedures than most other industry sectors, there were still substantial areas for transformation (Ligierko* 1994). As Krzyzanowski (1993) details, Polish shipping companies have operated under socialist law, based on crew self-government and on the financial management ethics associated with state enterprises. Authority within the companies was split between the management, employee's organisations and the trade unions, all of which resulted in substantial conflicts of interest and subsequent disbenefits for the organisations. The need to reorganise these business practices and restructure ownership to increase competitiveness, were a major incentive to privatise.

There are two main mechanisms of privatisation in Poland; transformation (or capital transfer) and liquidation.

Liquidation - the most common method used so far in Poland generally - it has been used primarily in small to medium sized enterprises, but is not yet relevant to shipping enterprises.

Transformation - (preferred by large organisations)

1. State-owned companies restructure themselves as limited liability companies or joint-stock companies which are wholly owned by the State Treasury (PZB remain at this stage). They then supposedly operate in accordance to a commercial code established by the Ministry of Privatisation in the past and now administered by the Treasury.
2. Actual privatisation of the newly structured organisation; via public sale, auction or Government Mass Privatisation Programme (Ledger and Roe 1993). This is not as easy as it sounds because as Sawiczewska (1992) indicates, "..there are very few people in Poland who have the money", and that it is estimated that the average staff member of a state owned enterprise, ".. would not be able to purchase shares for another 15 -20 years". In light of this there has been a political move to give shares in ownership to the employees at a very small cost which, though it is justified in light of the very poor wage of average Polish employees, does not raise revenues.

The Polish government is determined that all shipping and marine activities should eventually be run by the private sector (Dunlop 1992). Progress to date however, is generally slow and disordered. The Polish Act on the Privatisation of State Owned Enterprises (passed 1st August 1990), enabled the wholesale move to privatise hundreds of Poland's largest companies. Polish Ocean Lines and Polska Zegluga Morska were in the Ministry of Privatisation's original list of 400 companies to be privatised. Though legally restructured under the first stage of transition privatisation, both companies remain largely in the hands of the state. Zurek (1994), Ledger and Roe (1993), and Kujoth*(1994) amongst others, suggest that the process has been slow due to a number of factors. These include, the large size of the enterprises, the immense value of their assets, political instability, a public disillusionment with the reform programme, and a lack of disposable income amongst the majority of the population.

Many of these causes are beyond the control of the companies themselves and solutions will depend upon the developments within both Polish and world economic and political environments. However, there have been some signs of progress due to the restructuring of enterprises into holding and subsidiary companies. Here, smaller autonomous companies have been formed to carry out more specific lines of operation, as profit centres under a looser central control. This sectional approach to making privatisation more acceptable has been in evidence in several enterprises and is thought by industry authorities and as well as academics, to be the likely way forward (Ligierko*1994, Kujoth*1994, Dunlop 1992, Dobrowolski 1994).

Polish shipping under transition

The international nature of shipping ensures that external factors such as world economic climates, business conventions and international regulations, will have an effect on the conduct and performance of the industry. Concentration here however, will focus upon aspects of the industry more directly affected by those internal factors. Whilst privatisation in the shipping industry is progressing only slowly at the moment, the general advance towards it, has necessitated considerable structural changes and the introduction of market led business practices, in preparation for full privatisation. This has created an environment of great change which is likely to continue for some considerable time (Clayton 1994).

The loss of government subsidy has forced cost cutting measures upon managers. Some of these inevitably resulted in social costs such as unemployment, which in turn increased labour unrest and dissatisfaction (Ligierko*1994). Unnecessary administrative levels were eradicated as management themselves were restructured and the increase of computerised systems became apparent (Kujoth* 1994). Western advisors are widely utilised in the maritime sector, to ease the change, although, in comparison with other industry sectors, Polish shipping managers are relatively familiar with western conventions due to the international nature of the industry. The employment of seafarers and labour under the principles of cost efficiency has become a contentious issue, especially when the power of the trade unions is taken into consideration (Ligierko* 1994). The 1991 Labour on Merchant Sea Going Vessels Act allowed the employment of foreign officers and ratings on Polish ships provided that they held the appropriate qualifications (Dunlop, 1992) yet union pressure in an era of unprecedented unemployment has curbed this development.

A newly available option is that of flagging out Polish vessels to open registers, enabling enterprises to take advantage of economic, legislative and fiscal benefits (Chrzanowski 1979). This process is occurring in all parts of the Polish fleet. By 1995, over a third of the Polish fleet was flagged out, prompting the Ministry of Transport to consider the introduction of a Polish International register (Lloyd's Ship Manager 1995).

The current role of the shipping sector is firmly that of profit maximisation at best, or survival at worst. The reform programme of 1990 and its subsequent alterations brought about an era of high inflation (though lower than in 1989) which affected costs, and high interest rates - 53% in 1992 (Krzyzanowski 1993) and still around 25% in 1997. In addition, no tax advantages or protection measures against bankruptcy were afforded to the industry to ease it through the transition period. Subsidies were also reportedly eradicated and tonnage replacement in this climate has proved difficult (Zurek 1994). This has resulted in overall ageing of the fleet, a tendency towards uncompetitiveness in an already oversupplied world market and an increased need for foreign mortgage facilities. Zurek* (1994) identifies the need for greater governmental support through indirect measures such as credit guarantees and tax advantages, but not direct subsidies which would defeat the object of privatisation.

The Polish Baltic ferry industry

The basis of Polish passenger shipping was developed from the demand for labour emigration during the inter-war period. As the post-war industrialisation programme progressed, employment became freely available and, under an authoritarian regime, mass migration was curbed. Additionally, the cold war did not assist passenger movements and those journeys which did remain could be substituted by an increasingly affordable air service. The passenger liner sector consequently declined and remained in decline until the first ever Baltic ferry services began in 1965 (Chrzanowski 1977, Fallenbuchl 1980).

The initial service, between Swinoujscie and Ystad in Sweden, was operated by Polska Zegluga Morska (PZM). In 1970, under the administration's restructuring programme, ferry operations were transferred to Polish Ocean Lines where they stayed until 1976. New lines were added; 1973 Gdansk to Helsinki; 1974 Gdansk to Nymashamn in Sweden and Swinoujscie to Travemunde in the former East Germany, and; 1977 Swinoujscie to Copenhagen. The further fulfilment of the monopolistic, one sector-one company rationalisation process was completed when a new state owned operating company, Polska Zegluga Baltycka was formed in 1976. Polish Ocean Lines relinquished the ferry operations to concentrate on the liner trade and PZB took over as the sole operator.

Polska Zegluga Baltycka

PZB was based in Kolobrzeg primarily to aid the development of the central section of the Baltic seaboard and was to create a specialised organisation for ferry operations. Long term plans for the company included the activation of ports for commercial services, both ferry and cruise operations; hence it also became involved with tourist services such as hotels and restaurants (Chrzanowski et al 1979). This was unprecedented in socialist countries and proved to be troublesome during the future privatisation process (Zurek 1994). PZB was also involved in short-sea tramping services and operated the ports of Kolobrzeg and Darlowo.

Despite the monopoly position of the company, concerns were voiced regarding the problems which may be faced by Polish ferry operators. Fallenbuchl noted:

> The passenger traffic on the Baltic depends on tourists, particularly from Sweden, and it is very sensitive not only to political factors, the economic situation in the west, and the weather, but also to the capacity of the Polish tourist industry to absorb large numbers of foreign customers, and that capacity is growing very slowly (Fallenbuchl 1980).

The new company was formed by transferring four car/passenger ferries from POL and 19 small coasters from PZM. By 1988 the total fleet had fallen to 16 ships, though the ferries had increased in number to six, and, in the ensuing year, the fleet declined even further to 11 vessels through the non-replacement of aged tonnage (Ernst and Young 1990).

Though the company headquarters was based in Kolobrzeg, PZB operated from Swinoujscie in the west and Gdansk in the east, operating both of the ferry terminals. The main destinations during the 1980's were Ystad (Sweden), Copenhagen (Denmark), Helsinki (Finland) and Nymashamn (Sweden), though other services were tried on a seasonal basis and were rejected through insufficient traffic. The Swinoujscie-Ystad service, which was jointly run with POL who served the freight sector, was by far the most important ferry route; in 1988 it accounted for 54% of total passenger traffic and 69% of total passenger cars carried by PZB (Ernst and Young 1990).

In 1988, PZB was structured into 5 separate divisions; Ferry, Tramping, Port Handling, Tourism and Construction and Repair. Tourism was co-ordinated through city centre travel offices and the Ferry sections of the organisation. The company had autonomy to set fare prices but clearly indicated their chosen market by only quoting these in foreign hard currencies, a move to deliberately discriminate against the Poles, which caused a substantial amount of antipathy towards the enterprise (Kujoth* 1994 and Krasnicki* 1994).

Although the company enjoyed monopoly power for over a decade, PZB has had a rather "..chequered financial history" (Dunlop 1992). A grave financial crisis in the early part of the 1980s left the company on the verge of bankruptcy; it survived because of Government subsidies, the organisation of credit arrangements with PZM, and severe austerity measures. Because of this poor performance, the company had not been able to reinvest in new tonnage and the ageing of the fleet became a genuine cause for concern (Kujoth* 1994). Some confusion exists as to exactly how the company faltered; the company's financial reporting was neither transparent nor widely available and the exact level of government support was always unclear (Zurek* 1994). Certainly, the labour dissent and subsequent imposition of martial law in the early eighties, severely affected passenger traffic in the ferry sector. Additionally, it must be remembered that PZB was involved in other areas of tourism, as well as port operations and short-sea tramping. Dunlop (1992) reported that there was a long term general depression in the latter sector and that tonnage was being reduced, which may provide some indication as to the major problem area. PZB existed as the sole Polish ferry operator until 1991.

The Polish ferry industry in the post communist era

The initial transition stages in the ferry industry have been as slow as other areas of the shipping industry, but the monopoly has been broken. With the recent operational start up of another enterprise, the pace of change is likely to continue to quicken. The sector is constantly changing and is likely to do so for a considerable time (Porzycki* 1994). At the time of writing however (October 1996), there are three existing operators; one is state owned (PZB trading as Polferries), one Swedish owned - Stena - trading as Lion Ferries, and one jointly owned by two state companies - POL and PZM - trading as Unity Line. The former Corona Lines SA was a wholly owned joint stock subsidiary of KTG sa, and as such was the first fully privatised Polish ferry company. It ceased trading, however, in 1995. Whilst the sector as a whole is still dominated by PZB, the development of the present structure displays many of the characteristic changes associated with the transition from a command, to a market economy.

Passenger carryings have begun to stabilise overall, following the period immediately after the fall of communism; a period which has seen an immediate and substantial increase in trade. The situation for the most popular Swinoujscie - Ystad route is significant with regards to two very specific causal factors of a major fall in passenger throughput. The periods of labour unrest and martial law in 1981/2 and also, the fall of communism in 1989, whilst at the same time the increase in freight carriage continued largely unaffected. There is still a substantial underlying positive trend in passenger traffic and feasibility studies carried out by Zegluga Polska and Euroafrica indicate that there is a considerable growth potential (Krasnicki* 1994) mainly due to the potential transit traffic between wider Scandinavia and central and southern Europe. In addition, Sawiczewska

(1993) and Kujoth* (1994) amongst others, estimate that growth in the passenger sector will continue at between 2-3% for the foreseeable future.

We will now concentrate upon PZB, as this is the only ferry company to have experienced both the old and new Polish business environment.

Polska Zegluga Baltyka (PZB)

The origins of the company as the state owned monopoly ferry operator were outlined earlier. PZB was not one of the 400 state owned enterprises targeted for privatisation under the Balcerowicz mass privatisation programme (Zurek*1994). It still lies firmly within state hands, but there have been efforts to restructure the enterprise to enable eventual privatisation. Even though PZB was the first of the three major shipping companies to be fully converted into a joint stock company, it remains in a state of transition. The planned finalised structure will see PZB become a holding company with subsidiary companies operating in specialised fields - ferry services being one of them (Sawiczewska and Zurek 1994). The state is the sole shareholder which means that ultimate responsibility is still to the Polish Treasury, but supposedly the company receives no subsidy back from the state (Zurek* 1994).

Although PZB still dominates the industry it is thought that it will stay under state ownership for a considerable time. Little interest has been shown from the private sector. Kujoth (1994) explains:

> The company is still wholly owned by the government......because it is losing money and those with money to invest are therefore not interested. They would rather put it into other sectors, with better potential returns.

In 1994 PZB operated services on the following routes:

Swinoujscie -Ystad 14 trips/week
Swinoujscie -Copenhagen 5 trips/week
Gdansk - Oxelosund 2 trips/week
Gdansk - Helsinki 2 trips/week (now closed)
Swinoujscie - Ronne 1 trip/week (summer only)

These remained largely unchanged since the mid 1980s. Since the advent of democracy in Poland there has been a steady overall increase in passenger traffic between Poland and Scandinavian countries. PZB did, however, experience a fall in market share on the Poland to Sweden routes, due to competition from new market entrant, Corona Lines. By 1994 Corona Lines had captured 25% of the market (Cruise and Ferry Info. 1994). However, the failure of Corona Lines will clearly have some impact in the market, although this is likely to be compensated by the entry of Stena Lines, through Lion Ferries, and POL/PZM through Unity Line although they are trading on the Ystad - Swinoujscie route to the west of Poland.

PZB is still the only shipping company involved in port and terminal operations. In April 1994 the first stage of the redeveloped PZB ferry terminal at Swinoujscie opened. It is due to be fully operational by 1996. Annual capacity will reach 800,000 passengers, 150,000 cars, 100,000 trucks and 60,000 rail wagons. The estimated cost of redevelopment is approximately $76m. Perhaps surprisingly it was reportedly financed in part from the company's own funds and in part through state contributions, with no bank finance whatsoever.

PZB have not upgraded their ferry fleet for many years, and it now consists of five vessels ranging between 16 and 19 years old, which all sail under the Polish flag. The age of the fleet represents a major problem since replacement funds are reportedly difficult to find. This seems to be a grey area; a number of replacement vessels had been planned for the early 90s but did not materialise (Ernst and Young 1990, Dunlop 1992), though it is not clear whether this was due to financial reasons. Certainly the financial situation in the 1980s was poor, but the development of the ferry terminal suggests that finance is available, whether from company funds or from government subsidy, although subsidies were supposedly non-existent within the guidelines of the Ministry of Privatisation.

Assuming that finance restrictions will not permit vessel purchases, then chartering foreign vessels may provide a short term solution. PZB now charters 4 ships (1 ferry and 3 bulk) on bare boat charters from UNI-CHARTER, a Hamburg based company. This is only possible now because of the liberalisation of Polish shipping registration in 1988. Previously it was only possible to charter foreign tonnage through Polfracht, the state owned agent (Ernst and Young 1990).

In terms of operating strategy; PZB are reportedly concentrating on the ferry sector because of the general increase in ferry traffic and the slump in the Baltic short sea tramping sector (Dunlop 1992). Whilst the company is ultimately responsible to the Polish Treasury before full privatisation, it is directly subordinate to the remnants of the Ministry of Privatisation and as such, is subject to the rules governing operating procedures under this form of transition (Zurek 1994). Opinions differ as to the degree with which these operating procedures have been absorbed by PZB.

In this company, as in others, economic and financial conditions of the business are constantly modified, improved and adapted to the new market conditions. The changes in the organisation system are based on simplification of structures leading to full privatisation within the framework of a joint stock company (Krzyzanowski 1993).

Following the transition to a market economy all companies in the state sector in Poland are expected to operate on a fully commercial basis. PZB management has therefore been given complete independence in the daily operation of the company, and is also responsible for its performance and financial position (Dunlop 1992).

These positive statements are challenged by Kujoth* (1994), the former Assistant to the Managing Director at PZB, arguing that PZB were still operating as if they enjoyed a protected state monopoly position and that consequently, apathy was rife and that there was considerable inertia to change within the organisation. It was stated that by 1992 no changes had been made to the management practices, no new marketing strategies existed and no investment in IT had been made. Further criticisms were forthcoming regarding the company's isolationist attitude regarding possible inter-industry co-operation or mergers. This was confirmed by Ligierko* (1994), and Porzycki* (1994), with regard to the joint venture, Unity Lines.

In the context of marketing, no radical changes in strategy have become apparent. PZB's passenger marketing is still co-ordinated through city centre travel offices and the ferry sections of the organisation, although an additional office has opened in Kolobrzeg. The fare prices are still not quoted in zloties (PZB 1994).

As far as future developments go, it appears that unless one operates within the organisation, the position of PZB is difficult to assess accurately. Future strategies and developments cannot be commented upon with any degree of certainty, but there are a number of issues that have to be faced. These will arise from, a) within the company; such as the need to replace tonnage, improve the financial position, change management attitudes and complete the restructuring programme, b) within the newly structured industry; such as the need to face increasing competition, and, c) events outside the influence of the industry, which will be faced by all players.

Future considerations

Breitzman (1992) suggests that there are four broad factors which will determine the future of the passenger ferry sector in the Baltic area, namely:

1. The general increase of trade between countries which have direct access to a Baltic seaboard.
2. The dominant position of road transport in the movement of both passengers and goods.
3. The increasing mobility of people caused by increasing incomes and spare time.
4. The degree of competition from alternative modes.

The underlying trends appear to indicate that Breitzman's initial determinant is reflected in the Polish passenger sector. There are however, factors peculiar to Poland, which could well affect the future relative competitiveness of Polish operators to those from surrounding states.

With reference to the dominance of road transport. Poland is currently in need of a substantial improvement in infrastructure, but, if this occurs, the country is placed well geographically, to become a major cross-roads for European passenger (and freight) movements. Forecasts for trade are optimistic, for example:

> Transport between Scandinavia and Poland is forecast to grow by leaps and bounds - average annual daily traffic through the Karlskrona area is expected to reach approximately 2000-4500 vehicles in the next 25 years (Polish Maritime Industry Journal 1994).

But as Sawiczewska (1993) describes at some length, and Kujoth* (1994) confirms, it is reasonable to declare that the present road infrastructure in Poland is far from adequate to be able to satisfy the level of forecast demand. Optimism is found in the planned development of two major road schemes in particular; the Trans-European Motorway (North-South) and the Via Baltica, which feeds into it. The schemes are very capital intensive however, and the harsh economic realities faced in eastern Europe leave doubts as to whether such an ambitious investment programme is likely to be realised without considerable western funding (Lloyds List 1993).

With reference to increases in mobility, the availability of disposable incomes in Poland is important. However, if the general mobility of Europe as a whole increases, then central European countries can expect increases in both transit business and tourism within individual states. For Poland to be competitive in claiming the increased transit business, the aforementioned improvements to the road infrastructure will be of paramount importance, to compete with countries such as Germany (Sawiczewska* 1994). For ferry operators to benefit from increased tourism requires the ability to attract holiday makers to stay in Poland and, the ability to cope with the demand if they succeed. The tourist industry in Poland during communism was centrally planned, with a single travel agency (Orbis), and was engineered to serve other CMEA countrie's tourists rather than those from the West. In the mid-eighties Poland was 18th out of 21 European countries in terms of the number of tourists it received. With negligible interest on the part of major western travel companies and a relative lack of prominent attractions in comparison with other central European countries, the marketing effort will have to be sizeable (Dawson 1991).

The threat of new competition from alternative modes is most obviously manifested by the construction of the fixed road bridge and tunnel link between the continent and Sweden. However, because the route links Sweden to Denmark and Germany, it is thought that a greater effect will be felt by ferry operators in the extreme west of the Baltic, rather than by Polish companies (Ligierko* 1994).

Conclusions

Whilst it is true that PZB, as a state controlled enterprise, still dominates the ferry sector, there are sufficient indicative factors to suggest that the reform measures have influenced the structure, conduct and performance of the industry in the post-communist period. The first obvious effect of the fall of the previous regime was the immediate rise in passenger throughput, especially in the first two years of the 1990's. Here the perception of a generally more accessible eastern Europe is considered to have accounted for the increase in foreign travellers. Although this growth has levelled off to a certain extent, there is still an underlying positive trend, which is conservatively forecast to continue at a rate of approximately 2 - 3% per annum. This at least provides an encouraging context in which the sector can develop.

One of the fundamental constituents of the Balcerowicz plan was the mass privatisation of state owned companies. Within every sphere of Polish industry, progress towards this goal has been much slower than desired and this is certainly the case in the shipping industry. PZB remains wholly in state hands and all indications are that this will continue for some considerable time. There is little public willingness to invest in shipping generally, and especially in an unprofitable concern. The company however, has undergone a nominal restructuring programme and was the first of the major state owned companies to become a joint-stock company in preparation for privatisation. The ultimate intention is that PZB will eventually become a holding company with subsidiaries concentrating on well defined areas of operation. Now directly subordinate to the newly expanded Treasury since October 1996, it is supposedly acting according to the code of conduct established for transitional enterprises. There is contention from certain industry sources as to whether or not PZB has changed at all, despite the restructuring programme. Additionally, there are still indications of government subsidies (Lloyd's List 1994) though this is denied in some quarters (Zurek* 1994), and, admittedly, this area is unclear.

The movement of the other shipping companies towards privatisation is reflected within the ferry sector. The sectoral approach to privatisation as adopted by both POL and PZM has led to the formation of subsidiary companies. The need to offer potential investors smaller, more streamlined, cost conscious enterprises, led to the creation of Euroafrica (POL) and Zegluga Polska sa (PZM). Both of these subsidiaries have begun to make a significant future contribution to the ferry sector through the joint-venture, Unity Lines. The fact that PZM and POL are allowed to re-enter the ferry sector reflects that the functional demarcation of the shipping industry by the communist regime has been eased. Through the example of Unity Lines and Lion Ferries it can be seen that the structure of the industry is now determined by the market rather than by the state.

Arguably the most significant development in the ferry sector has come about because of legislation passed before the initiation of the Balcerowicz measures. The last pre-Solidarity government had begun to lift the restriction on the creation and development of private firms in 1988. By 1993 over half of Poland's GDP was produced by the private sector, predominantly, it is thought, due to the growth of these new firms and not the privatisation of state enterprises (Lloyd's List 1994). Corona Lines' parent company, KTG sa, was established in the period immediately after these deregulatory measures became law. After a faltering start, Corona Lines became the first ferry operator to successfully achieve a break in the state monopoly of the sector. In only three years, the company captured 25% of the market share for services between Poland and Sweden. However, its failure in 1995, stemming from an inability to cover ship mortgage repayments from revenue after paying other start-up costs, suggests that a full market structure for the ferry industry in Poland remains some way off. Its replacement as a private competitor by Lion Ferries provides further development in this sector with the sizeable and strong backing of the Stena organisation.

Corona Lines began operations with a heavy western influence on the managerial and operational techniques which it employed. There was a recognition that other companies could enter the market just as they had done and that the threat posed by this potential competition required the company to operate as efficiently as possible. The Swedish finance and expertise employed from the outset meant that the company's conduct could be moulded to meet market pressures from day one of its operation. The adoption of western expertise and cost efficiency measures has also been recognised as essential to the success of the joint-venture, Unity Lines. However, because of the origins of the participants, some of the efficiency measures proposed, such as the reduction of manpower are likely to cause problems with the parent companies, from the incumbent trade unions. Corona Lines operated a "no-union" policy with regard to staff for this very reason, and, have also invested heavily in information technology to minimise administrative salary costs. The era of total employment in Poland is over as the demands of the market and the reduction of protectionism mean that unnecessary staff costs need to be curbed. The challenge to PZB to adopt competitive practices in response to the appearance of the new operators has been either met, or alternatively ignored, according to who is to be believed.

The 1988 liberalisation of Polish shipping registration allowed operators to minimise costs further by flagging out existing vessels or chartering foreign flagged vessels. Both Corona Lines and Unity Lines used and use respectively foreign flagged vessels exclusively, although much uncertainty surrounds PZB. In the latter case it was necessary to flag out the newbuilding to secure a foreign mortgage, a matter of compulsion since current interest rates in Poland are

77

prohibitive to large scale investments - a matter of significance here since the high level of interest rates were determined under the IMF approved transitional measures.

The new era of competition has witnessed the need for effective marketing and heightened market awareness amongst operators and again here, western influences are evident within the sector. There is a move towards the sale of ferry travel as a product in itself, rather than the traditional view of the service being merely a means to facilitate travel between origin and destination. Both Corona Lines and Unity Lines, and now Lion Ferries, in contrast to PZB, recognised the need for a strong brand identity and an emphasis on service quality to shift public opinion away from the unfavourable attitude, historically given towards Polish ferry services. Improvements to service levels was attempted by the vetting and training of "Hotel" staff by an overseas agency in Corona Lines' case and international hotel chains in the case of Unity Lines. In the latter case the quality of the newbuilding has also been emphasised in the marketing campaign. There are many challenges in this area, not least of which is the surfeit of high quality competition in the Baltic Sea from Scandinavian operators.

It is recognised that predictions with regard to the shipping industry have traditionally left much to be desired (Ledger and Roe 1993). In the case of the Polish ferry sector, the fact that the economic, political and social changes within eastern Europe are unprecedented, adds a further caveat to the validity of attempts to predict future progress within the sector. It is however, true to say, that fundamental changes have already been observed, in that the element of competition has been introduced. This in turn has demanded the westernisation of managerial and operational strategies. These have received a broadly positive response by the new entrants; there is considerable doubt regarding PZB in this matter. The attempts to develop are not always helped by the prevailing economic and political uncertainties within Poland or the government's laissez-faire attitude towards the transformation of the shipping industry. Moreby's (1989) statement has much relevance here:

> For capitalism read free market forces; for communism read central planning; and for confusionism, read a mixture of the two.....Today, shipping is dominated by confusionism - so many players wanting to operate free-market forces while, at the same time, wanting some degree of help and support from governmental central planning (Moreby, 1989).

References

Breitzman K.H. 1992 Ferry transport in the Southern Baltic sea and its prospects, in Wijnolst N., Peeters C. and Liebman P. (eds) *European Short Sea Shipping* Lloyds of London Press Ltd, London.

Chrzanowski I. 1977 Polish shipping and shipping policy. *Maritime Policy and Management*, Vol 4, No 5. pp281-292.

Chrzanowski I., Krzyzanowski M.T. and Luks K. 1979 *Shipping Economics and Policy: A Socialist View*, Fairplay Publications, London.

Clayton R. 1994 Working without Warsaw: World in focus; Poland. *Fairplay*, 12th May.

Cruise and Ferry Info. 1994 Annual traffic statistics, March.

Dawson A.H. 1991 Poland, in, Hall D.R. (ed) *Tourism and Economic development*. Bellhaven Press, London.

Dobrowlski K.1994 The privatising of state owned enterprises in Poland's economy, in Zurek J. (ed) *Processes of Changes in Ownership in Maritime Transport in Poland and East Germany*. University of Gdansk, Poland.

Dunlop A. (ed) 1992 *The Polish shipping and maritime industries*. Charter Newsletter Ltd, Oxford.

Ernst and Young 1990 *Analysis of the Maritime Transport Sector of Eastern European Countries, Final Report: Volume 4, The Maritime Transport Sector in Poland*. Commission of the European Communities.

Fallenbuchl Z.M. 1980 Poland's maritime transport, in Mieczkowski B. (ed),. *East European Transport: Regions and Modes*, Martinus Nijhoff Publishers: The Hague.

Krasnicki* A. 1994 Personal interview between Neal Toy and Dariusz Gluszczak, Area Manager Poland for Corona Lines. 17th September 1994.

Krzyzanowski M.J. 1993 The adaption of the Polish shipping companies to market conditions, *Maritime Policy and Management*, Vol 20, No 4, Oct - Dec.

Kujoth* A. 1994 Personal interview between Neal Toy and Andrej Kujoth, Managing Director of Corona Lines, (previously PZB) 15th September.

Ledger G. and Roe M. 1993 East European Shipping and Economic Change: a Conceptual model. *Maritime Policy and Management*, 1993, Vol 20, No 3, pp 229-242.

Ligierko* Z. 1994 Personal interview between Neal Toy and Zbigniew Ligierko, Deputy Managing Director of Euroafrica shipping Lines, 17th September.

Lloyds List 1994, Poland. Lloyd 's List, 27th May 1994.

Lloyds Ship Manager 1995 Poland plans open register, March.

Moreby D.H. 1989, Capitalism, Communism and Confusionism: the three ideologies affecting modern shipping, *Maritime Policy and Management* (1989), Vol 16, No I, Jan - Mar.

Patton M.Q. 1987 *How to use qualitative methods in evaluation* Sage Publications, London.

Polish Maritime Industry Journal 1994 Port of Gdynia campaigns for new TEM-NS terminal. Polish Maritime Industrial Journal, No 2, August.

Porzycki* P. (1994) Personal interview between Neal Toy and Pawel Porzycki, Managing Director of Unity Lines, 17th September.

PZB 1994 Polferries timetable and prices, Summer, unpublished.

Sawiczewska Z. 1992 Reconstructing Polish ports and shipping, *Maritime Policy and Management* Vol. 19, No.1 pp 69-76.

Sawiczewska Z. 1993 The impact of political and economic change on Polish short sea shipping, in Wijnolst N., Peeters C. and Liebman P. (eds) *European Short sea Shipping*. Lloyds of London Press Ltd, London.

Sawiczewska Z. 1994 Personal interview between Neal Toy and Professor dr hab. Zofia Sawicjewska, University of Gdansk, 14th September 1994.

Zurek J. 1994 Privatisation of Polish shipping enterprises: Present situation and directions of development, in Zurek J. (ed) *Processes of Changes in Ownership in Maritime Transport in Poland and East Germany*. University of Gdansk, Poland.

Zurek* J. 1994, Personal interview with Professor dr hab. Janusz Zurek, University of Gdansk, 14th September 1994.

* Personal interview conducted by Neal Toy.

The North South motorway and its impact upon activation of Polish ferry shipping

Janusz Zurek
Institute of Maritime Transport and Seaborne Trade
University of Gdansk

Abstract

This paper discusses the new North-South motorway and its impact upon Polish ferries. It looks at the history of Polish motorway construction and the current plans for development in the coming years, with particular reference to the relationship between the A1 motorway and maritime investments.

1. The motorways as the basic elements of the country's infrastructure - the programme and construction terms

In the post war period in Poland, only a few short stretches of motorways were built. Nowadays, with a view to making up the losses in that field, the construction of motorways is being planned - at a rate of 150 - 200 km annually. The programme of motorway construction underlines that in the next 15 years almost 2.6 thousand km of motorway will be completed. The total value of the whole project is preliminarily estimated at US$8-10 million. The first tolled motorway, linking Katowice and Krakow, will be in operation in two years.

The law of October 27 1994, came into force on January 21 1995. This Act along with the associated decrees, constitutes the legal base for realisation of the programme for motorway construction. It defines both the conditions for preparing the contracts for building and exploitation of the toll motorways, and the rules of gaining the licences for it. Construction of motorways is not only an economic investment but also a social and political one - of great value for the economic development of the country. The network of motorways is an essential element of integration, through a modern transport system, into international cooperation in Europe. That fact is of special importance in the light of future Polish membership of the European Union.

The Act, which underlies those changes, adapts the rules and norms in the transport domain to those which are valid on the territory of the European Union countries. The reason for these adjustments arises from the Agreement on Association of Poland with the European Community, of 16 December 1991. This Agreement concerns three basic fields: uniformity of legislation, modernisation and development of transport infrastructure, transformation in the field of proprietorship and organisation, and changes in the management of the domestic transport enterprises.

According to the programme of motorway construction, four toll motorways are expected to be built in Poland in the next 15 years. They are the following: A-1 Gdansk-Torun-Lodz-Czestochowa-Katowice-national border with Slovakia: A-2 Swiecko-Poznan-Warszawa-Terespol-national border with Byelorussia: A-3 Szczecin-Zielona Gora-Legnica-national border with Czech Republic: A-4 Zgorzele Wroclaw-Katowice-Krakow-Tarnow-Rzeszow-national border with Ukraine.

The Agency for Motorways Construction and Operation has been established. by the state and is obliged to prepare the basis for the location of routes of the above mentioned motorways, and to purchase and safeguard land for the planned motorways. It should also prepare allocation by tender for construction and operation of the separate stretches of the motorways. The procedure of allocation by tender may be entered into only by limited liability and joint stock companies with a base in Poland, with an initial capital not less than 10 million ECU. Gaining a licence puts the licensee under several obligations: to accumulate capital for building and exploiting the motorway, to prepare technical specifications - necessary for beginning motorway construction, to obey the technical, construction and operational rules which apply to the motorways, and to obey the regulations concerning protection of environment and cultural values, to conduct the investment process in due time, and finally to maintain the motorways in compliance with the standards of the law and licence agreement.

Realisation of such a wide programme requires explicit financing sources. At present, construction of short stages of motorways and infrastructure is financed by loans from international financial institutions. This clearly restricts the speedy realisation of such a large scale programme. Hence, the larger parts of motorways will be built, as tolled motorways, by the economic units which will gain the licence by tender. This idea is central as a strategic assumption of the construction programme. The chief point of such financing whilst building motorways lies in relieving the national budget by involving private capital. In order to finance the motorways, a number of consortia have been established, e.g. Polish and foreign capital companies acting as investors. These consortia satisfy the requirements of the licences for building and exploiting the motorways. The Treasury will bear the expenses of buying the land and expenses connected with the preparatory works - which will amount to 15% of the total costs. The land will constitute the State share in the motorways programme. The licensees will pay special fees for land utilisation which will be agreed in a contract.

The motorway construction will create possibilities to utilise the productive capacity of Polish enterprises, materials, machines and labour. It is predicted that each km of motorway will create working places for over 700 employees e.g. workers employed in building and other related enterprises, in institutions and organisations offering services for travellers. As a consequence, it will benefit the economic growth of the country as a whole.

2. The A-1 motorway as an important element of activating the maritime economy on the Eastern Coast

The European Commission and the UNO have determined that the North-South Transport Corridor should be in the most eastern location, passing from Finland, via the Baltic Sea, Gdansk region and further to the South - to the Mediterranean Sea, and via Istanbul to the Arabian countries of the Near East - to the Persian Gulf. Within the bounds of this corridor, the EU and UNO have agreed construction of the Trans-European North-South Motorway - the so called TEM. This motorway runs from Gdansk, through Poland, former Czechoslovakia, Hungary, former Yugoslavia, Bulgaria, Turkey and Arabian countries of the Near East to the Persian Gulf.

The Eastern region has special interest in building this motorway as one of the first. It is motivated by the benefits which will result from its specific location, by the existence already of part of the motorway near Piotrkow Trybunalski, by a bridge over the Vistula river which is under construction, and moreover by the prospects for development of trade between Scandinavia and the Balkans. The Gdansk Motorway Consortium has established a company which will compete for a licence for building 334km of the A-1 motorway, from Gdansk to Tuszyn (near Lodz). The following enterprises are members of the Consortium: The Bank of Gdansk, refinery of Gdansk, Gdansk Port Authority and the Commercial Port of Gdynia.

Thanks to the motorway, an important economic hinterland will become open to Gdansk and other Polish ports. Thus new possibilities for transit and passenger transportation will be created.

The total length of the whole motorway will amount ca. 10,000km; 4,000km has already been completed, 3,500km are under construction and 2,500km are in a project stage. Unfortunately, Poland, which was once the initiator of the planned motorway investment, nowadays occupies the lowest place among the executors. It is not only economic reasons that have contributed to this, but also the political situation in the eighties and different contradictory interests of local authorities.

The Port of Gdansk is planned to be the initial motorway junction for sea and land transportation. The port is expected to be developed to the east, beginning in the old port of Gdansk where the port still has some reserves of undeveloped land. The modern port terminal of the A-1 motorway (planned for passenger car and general cargo) will cover around 5sqkm - with access to the port waterfront of 2.5km length situated on the Southern Bank of the Dead Vistula, near the Wislinka. The terminal is planned to have 12 berths, mainly for ro ro vessels and ferries. Each berth requires 200m width - with access to the port waters. Locating the main terminal here will avoid access via the main Gdansk city districts (when using the motorways). It is planned that the new Gdansk terminal, situated in the eastern Gdansk Port which is linked directly with the A-1 motorway, will handle yearly around 2.2million tons of cargo transported via the ro ro system.

The central terminal is expected to be the main junction of combined transport on the North-South Motorway. Development of the port complex, linked with the A-1 motorway, will assure, beside the million tons of general ro ro cargo shipped to Gdansk, Finland and other Scandinavian ports, an increase of tourist traffic - which could reach 5-6 million foreign tourists annually. The estimated annual income from the services offered to the tourists could amount to 3-4 million US$. It will encourage construction of a modern tourist district near the terminal, providing all kinds of services for foreign tourists, mainly those in transit. The area of the tourist district should be adjacent to the terminal. The A-1 motorway will also serve the port of Gdynia - thanks to the almost completed new branch.

With a view to the above mentioned factors, one should remember competition with the neighbouring Baltic countries, which could be interested in taking over at least part of this income (in foreign currency). Competitors could include the port of Tallinn - situated near the port of Helsinki. Tallinn and Helsinki are already linked by passenger car ferry lines. Estonia plans to build a motorway from the Tallinn port, through Estonia, Latvia and Lithuania - and going on to reach Warsaw and Lodz linking with the Trans-European North South Motorway (A-1) and thus the hinterland of Polish ports.

Construction of the A-1 motorway is important to Poland due to the fact that it will enable better utilisation of the seaside region of the country and the revival of the maritime economy in the eastern regions. The motorway will attract onto the land routes valuable general cargoes which were bring transported by sea. It is hardly to be wondered at, that in such a situation several western port groups have undertaken action which does not contribute to the construction of the discussed motorway. They strongly push the idea of building motorways from east to west which will enable a major part of cargoes to be directed to the west European ports - to Hamburg, Antwerp and Rotterdam. It is expected that this East-West motorway will be built using mainly foreign capital

The A-1 motorway is especially in the Polish interest because it will shorten the time of transportation to the Mediterranean seaports from 10-14 days (using sea transport) to 3 days (by land transport). Thus the Trans-European North-South Motorway will cause serious changes on the regular sailing lines. Some ship owners declare a division of regular lines into feeder ones - from the Baltic ports to Gdansk/Gdynia and on the Mediterranean Sea to the ports situated on the ends of motorway branches and to Istanbul (and vice versa).

Among the first changes of that kind on the Baltic, one could include establishment of the Swe-No-Po organisation which joins representatives of Sweden, Norway and Poland. Swe-No-Po patronises the opening of the shipping-ferry line from Karlskrona to Gdansk/Gdynia - linked with the A-1 motorway. Due to the lack of a modern passenger-car terminal in Gdansk, today's operations take place at the ferry base of the Polish Baltic Shipping Company which can handle only one ferry. That base which used to serve the ferry line Gdansk-Helsinki, provides the possibility of transporting cargo on a ro ro system from Gdansk to Iran, Iraq, Kuwait, Saudi Arabia, and on the return journey - using some parts of the A-1 motorway which have been completed and using replacement roads where the motorway is incomplete. Such roads in Poland have insufficient traffic capacity, hence the construction of motorway between Gdansk and Tczew is of strategic importance.

It should be mentioned that construction of the A-1 motorway in Poland will be highly complementary to the A-3 and A-4 motorways. These roads are planned to link Polish towns situated in the hinterland with the seaports. Thus it facilitates transit traffic from Byelorussia and the Ukraine to Gdansk and other Polish seaports.

3. Polish ferry shipping - with special regard to the impact of the A-1 motorway

The context for Polish ferry shipping does not look optimistic. In comparison with shipping in other Baltic countries, Poland's is of a very low standard - with respect to the condition of ferries (the oldest ones on the Baltic Sea), the location and condition of ferry terminals and their low traffic capacity which does not meet demand, and the under-developed transport infrastructure in the ports' hinterland. It puts shipping in a disadvantageous situation in the face of growing competition and it results in the loss of the position which should be maintained by Poland as a country situated at the crossing of the Trans-European transit routes, both North-South and East-West.

Further development of Polish ferry shipping depends on several factors. Replacement of worn out tonnage, which is becoming less and less competitive, is one of the basic conditions for further successful operation. The Polish Baltic Steamship Co (PZB) - still the leading Polish ferry ship owner - needs to introduce in the near future at least two new ferries - allocated to fixed routes. Fulfilment of such a plan is at present impossible. A typical ferry of 600 passenger places, with a suitable deck for cars, costs from 70 to 100 million US$ - and that sum surpasses the financial capabilities of the shipowner.

Table 1
The present situation of the Polish ferry fleet 1995

Ferries	Year of construction	Tonnage GRT	Shipowner
Rail-car 'Jan Sniadecki'	1988	14,417	Euroafrica
'Mikolaj Kopernik'	1974	2,988	Euroafrica
Passenger-car 'Silesia'	1979	7,414	Polish Baltic Steamship
'Pomerania'	1978	7,414	Polish Baltic Steamship
'Rogalin'	1974	4,020	Polish Baltic Steamship
'Nieborow'	1973	5,823	Polish Baltic Steamship
'Wilanow'	1966	4,020	Polish Baltic Steamship

Some important developments in Polish ferry shipping occurred at the end of 1993. An investment initiative was undertaken by the Unity Line Company - established by Euroafrica (POL), Zegluga Polska SA (PZM) and using Norwegian capital sources. The company signed a contract with a Norwegian shipyard in Tomrefiord to build a modern rail-car-passenger ferry - 'Polonia' - of 30,276

GRT. The ferry was launched in the middle of 1993, and now sails between Swinoujscie and Ystad transporting carriages, ro ro cargo, cars and up to 912 passengers. The ferry is the most modern operated by Polish shipowners - equipped with the latest facilities in the field of passenger safety, and all modern conveniences. Polish Baltic Steamship (PZB) refused to participate in the investment despite having the opportunity to do so. In turn, PZB as a shipowner has opened a new line from Swinoujscie to Malmo - which is situated only 45km away from Ystad. It creates a new era of competition between these two Polish ferry shipowners who operate on the Baltic Sea. Such intense and local competition may lead to economic difficulties particularly in the context of an increasingly cut throat market place with the new services operated by Lion Ferries and their commercially strong backing of Stena Line from Sweden.

The ferry terminals in Polish ports constitute important elements influencing the future of ferry shipping. Construction of a new ferry terminal in Swinoujscie will have a favourable impact upon services for both passenger and cargo traffic. The base serves 180,000 passengers, 50,000 cars, 100,000 trucks, 60,000 carriages and 2,620,000 tons of cargo each year. The new terminal's facilities far exceed the total capacity of the previously existing base. Unfortunately the ferry terminals in Gdansk and Gdynia do not present the same standard of quality of services. The terminal in Gdynia is a temporary berth, separated from, but contained within, the container terminal, handling ferries sailing on the Gdynia-Karlskrona route. In addition, the ferry terminal in Gdansk has not been modernised for many years and the technical condition of the base does not allow the handling of larger ferries. Meanwhile, the network of connections between ports and the transport hinterland of Poland remains highly inefficient, and although now under planned improvement, still presents a serious disadvantage for all ferry bases.

The problem of connections with ports,which pass through the city centres, should be resolved in the near future. The E. Kwiatkowski route, which has been under construction for many years, will link the port of Gdansk with the city peripheral motorway.

Hopes for the development of ferry shipping and the transport system in the hinterland are growing due to the layout of the North-South transport corridor (by UNO) - and the A-1 motorway. The A-1 is expected to bring Polish shipping several benefits, such as: improvement of car transportation, facilitating growth in motor tourist traffic (both international and domestic), attracting large transit cargo volumes from the south of Europe, contributing to an increase in maritime passenger traffic and development of services for tourists in the coastal regions.

The A-1 motorway has proved to be an important element in the development of Polish ferry shipping, by providing cargoes and greater numbers of passengers. It will require a modern port terminal which has to meet substantial technical and organisational demands. The central port terminal of the A-1 motorway, situated in the eastern port of Gdansk, will be the first terminal in central east Europe adapted to handling container traffic within a ro ro system. Some action in terms of new tonnage investment in ferry shipping by Polish shipowners, will be necessary. Ships providing high quality services, competitive with other tonnage of Baltic shipowners, should be introduced. As a result, the A-1 motorway, when completed, should prove to be a significant chance for improving both Polish shipping and the Polish economy.

Restructuring and privatising Poland's seaports

Leopold Kuzma
Institute of Maritime Transport and Seaborne Trade
University of Gdansk

Abstract

This paper examines the restructuring and privatising of Poland's ports starting from the subjective-organisational analysis of the situation prior to the political, social and economic changes and going on to examine the impacts subsequent to the developments.

To present the changes taking place in Poland's seaports arising from changes in the political system, at least a brief presentation should be given of the management and development of the subjective-organisational structures of port enterprises in Polish seaports during the period prior to the changes.

The subjective-organisational structure of Poland's seaports in the years immediately following the war, was similar to that of the port of Gdynia in the pre-war period. The port's assets were the property of the State (Treasury) and administered by the Harbour Boards, but all services afforded on behalf of the ship - cargo handling, stowage, trimming, storage, loading, discharging - as well as forwarding, brokerage, expertise and control, and ship supplies were handled by numerous private enterprises competing with each other. The majority of these enterprises had reactivated their pre-war services, but some were newly established.

Co-operative and private enterprises were established alongside private port enterprises. The state owned included the Coal Sales Centre which took care of the handling and storage of coal, and "Paged" - engaged in the handling and storage of wood and similar cargoes. This was the period of three sector management of Poland's seaports - the private, co-operative and state owned sectors.

In the years 1948-49, the state quite clearly restricted the activities of private enterprises, including closing them down. It was often the case that private port service enterprises were transformed into co-operative enterprises, then state owned. The conclusion of this procedure was the resolution of Council of Ministers Economic Committee which resulted in the creation of Port Authorities which were established in each of the particular ports.

These Authorities took over from the Harbour Boards, the technical infrastructure of the port, technical equipment, as well as dockers and other employees, from all the port service enterprises, and became centralised monopolistic enterprises performing the functions of administrators of the infrastructure and port terrains, as well as operational activities.

The process of eradicating private capital from the sphere of port services, also included forwarding, brokers, experts and control services, as well as ship supplies. Monopolistic state owned enterprises were also established in these fields.

Thus, at the beginning of the 1950s, the organisational model of the ports was based on the principles of:

1   All port services would be handled by state owned enterprises.
2   The subjective separation of technical services on behalf of the cargo and ship, from dispatching services i.e. forwarding, shipping and broking.
3   The concentration on a single level of each of these kinds of activities in the control of a single enterprise, maintaining the complementary character of production.

This subjective-organisational structure of Poland's seaports lasted to the end of 1991, despite the fact that it was the subject of numerous critical discussions and publications.

The centralisation of management of both infrastructure and operations inherent in such enterprises as the Commercial Seaports of Gdansk and Gdynia, as well as the Szczecin-Swinoujscie Port Authority,, gave rise to considerable negligence in respect of the maintenance and development of the infrastructure. The necessity thus arose to carry out changes in the ineffective, monopolistic system of port turnover, changes leading towards the establishing of a multi-direction system in each of the ports mentioned, by restructuring and privatisation in which competition will give rise to effectiveness of operations, to the advantage of the ports, shippers and carriers.

The changes in the ownership and subjective-organisational structures taking place in Poland's economy, including seaports, were directed towards the model provided by port operation and management in a market economy. The process of the changes discussed is complex and difficult, as it is taking place during a period when there is a lack of certain legal regulations, as well as a capital market. However, such a market is gradually taking shape.

Several steps have been taken in the ports, their aim being to change over to a market economy as smoothly as possible, by adapting the subjective-organisational structures and system of management to the new economic model. With this in mind the procedures needed to restructure and privatise the port economy have been worked out. These foresee three stages of transformation of the subjective-organisational and ownership structures in the field of port turnover:

> Stage 1 covers the conversion of the state owned enterprises: Gdansk Commercial Seaport, Gdynia Commercial Seaport and Szczecin-Swinoujscie Port Authority, into joint stock companies with the state as the shareholder. By late 1991, the first stage of this process had been achieved.

> Stage 2 covers the reorganisation within joint stock companies with the state as the sole shareholder. This had been largely achieved by October 1996.

> Stage 3 covers the removal of the Port Authority from operational activities and the privatisation of operating enterprises in ports. This was about to be commenced in earnest from November 1996 onwards.

At the end of 1991, the first stage was completed in all three ports, the place of state owned enterprises being taken over by newly appointed joint stock companies with the state as sole shareholder, viz. Gdansk Commercial Seaport, Gdynia Commercial Seaport, and Szczecin-Swinoujscie Port Authority.

At this stage, Supervisory Councils of the Joint stock companies and Company Boards, were appointed. In each port, the head of the Board is a Chairman elected by the Port Employees Council.

May 1991 was the last month in which Gdansk Commercial Seaport operated as a state owned enterprise. The process of transforming Gdansk Commercial Seaport into a joint stock company had two main aims: the separation of the port's operational-technical function from that of management and administration of the port terrains and infrastructure, and also the establishment of independent economic enterprises engaged in operational, technical and service-auxiliary port services which emerged from organisational centres of former state owned enterprises, and to compel port operation according to market rules.

In July 1991, the restructuring of enterprises was carried out in Gdansk Commercial Seaport Joint Stock Co.. Operation was isolated from the activities of the enterprise, by the establishment of 30 limited companies as employees' companies (only port employees could be shareholders). As the nominal capital of the limited liability company was not very high, Gdansk Commercial Seaport contributed 45% of the shares to strengthen it. In this situation, Gdansk Commercial Seaport Joint Stock Co. fulfilled the function of the mother company, the fixed capital of which (infra and supra structure) was handed over for the use of the limited liability companies.

The limited liability companies paid the mother company rent for the use of warehouses and storage yards, depending upon the area and for the use of quays - depending upon the length of the section of quay used. As regards the use of cranes and mechanical appliances, the companies recovered only the depreciation costs.

This organisational status of the Commercial Seaport Joint Stock Company under the joint board which continued to manage the infrastructure and operating of the 30 limited liability companies, survived to the end of November 1993. During this period, the legal regulation of the situation of the ports was awaited, namely the passing of the law on ports, by the Parliament, which was to settle the matter of changes of ownership and the privatising of enterprises in the sphere of port services.

The dissolution of the Parliament and Senate by the President, elections to both, the appointment of a new government and the activities of the Szczecin lobby in the matter of changes in the draft law on ports, resulted in a delay in its being outlined until October 1996 with a projected passage into law early in 1997.

In these circumstances, despite the lack of legal regulations, the Supervisory Board of the Commercial Seaport Joint Stock Co. and Company Board decided to separate the functions of the board from that of operations. Adopting an informal, but efficient method, Gdansk Port Authority was called into being, with responsibility only for the management of the infrastructure (development and maintenance) and the administration of the port terrains. The Port Authority also sold to the individual workers' companies, with limited liability, the previously leased supra-structure with a total value of 400 milliard (old) zloties, also selling their shares in the companies. The Port Authority Joint Stock Company continues to charge for the lease of terrains and warehouses. Payment for supra-structure sold to companies was broken down into instalments over a period of 10 years at low interest rates.

To promote companies engaged in operating the port of Gdansk, and to encourage cargoes to the port, "Gdansk Port Operating Co. Ltd" was established and maintained by the operating companies noted above.

The privatisation processes continue, irrespective of the restructuring of the Gdansk Commercial Seaport Joint Stock Company, new private enterprises being established apart form this. Thus two newly established handling stowage companies are operating. The first and largest - "Poseidon" - which employs a maximum of 2,500 temporary workers (these included in the firm's employment register), and the second firm - "Nawalcoop" - employing about 200 dockers. There is also a new company "Pilot Gdansk" offering pilotage services by sea captains who are qualified pilots.

Thus the structural ownership transformation and privatisation processes are most advanced in the port of Gdansk, and until such time as shares of Gdansk Port Authority Joint Stock Company are issued, the role of General Meeting of Shareholders has been held temporarily under the auspice of the Ministry of Privatisation and currently under the control of the Treasury.

According to the promoters of transformation in the port of Gdansk, the purpose of the establishment of the Port Authority Joint Stock Company and the other undertakings mentioned above was to establish a certain base situation which will in time enable further modelling of those element which require this.

In order to carry out the restructuring of the Gdynia Commercial Seaport Joint Stock Company, a special group was established, its task being to work out a general transformation programme. This included the group carrying out the division of the port's activities into three basic fields: management, operation and technical services.

As regards the authority, it was foreseen to transform the Gdynia Commercial Seaport Joint Stock Company into Gdynia Port Authority Joint Stock Company.

In the second sphere of port services, it was initially decided to set up 9 companies on five terminals - container, conventional general cargo, bulk cargoes, liquid fuel and grain, and four companies offering auxiliary services (ship services, transport, operating of handling appliances and mechanised gear).

In the third sphere it was initially decided to establish five technical enterprises - viz. - repairs to handling gear, mechanised equipment, buildings, water-sewage installations and electrical installations.

To determine whether and to what extent the planned service, operational and technical companies could function and bring in a profit independently, a simulation of costs, incomes and profits was carried out based on data from the previous year. The results suggested that the division into the service, auxiliary and technical enterprises was carried out both efficiently and effectively.

The lack of legal framework concerning the seaports held up the management of

Gdynia Commercial Seaport Joint Stock Company, from taking further steps towards introducing other changes within the company. In addition to this, the deliberate process of observing the introduction of these processes in other ports, and particularly in Gdansk, slowed down the Gdynia Company's further transformation.

The management of Gdynia Commercial Seaport Joint Stock Company, is convinced that the operating enterprises separated from the company should be small enough to be able to manage and adapt to changes, but large and strong enough as regards capital to guarantee their continued existence, resist external competition and attract both Polish and foreign capital.

The plan of the management of Gdynia Commercial Seaport Co Ltd foresees the establishment of a mother company, which would guarantee economic security for the companies isolated from the port with 100% capital of the Gdynia Commercial Port Joint Stock Company.

At the end of December 1993, the deed was signed which established the Baltic Container Terminal Co. (segregated from Gdynia Commercial Seaport Joint Stock Company). Meanwhile, work has already been completed to separate from the Gdynia Commercial Seaport Joint Stock Company, an independent unit with 100% shares owned by the port - this would be the Baltic Grain Terminal. Both terminals commenced work in the first half of 1994. The General (conventional and Bulk Cargo Terminals commenced work during the second half of that year.

The establishment of operating companies in which 100% of the capital is in the hands of the port, is a transition phase leading to privatisation proper, in which shares will be made available to the employees and a substantial number offered to both Polish and foreign investors in order to extend the company's activities

In the Szczecin-Swinoujscie Commercial Sea Port Joint Stock Company, there are 17 companies - enterprises with limited liabilities, eight of these being port services, the remainder being auxiliary and technical. The Seaport Company's share in the companies is 45%.

Further restructuring processes in the port group, have been held up due to the lack of a law on ports, as well the result of the demands of the Szczecin lobby that any law on ports should foresee the possibility of the establishment for this port complex of a Port Authority of Szczecin-Swinoujscie as an autonomous organisation.

These suggestions were included in the last draft of the law on ports, which may have accelerated its passage by Parliament. It turned out, however, that a new obstacle arose, namely that the local authorities and management of enterprises in Swinoujscie submitted a proposal to break away from the Szczecin-Swinoujscie Port Authority and establish their own separate Swinoujscie Port Authority. This new concept required further amendment to the draft of the law on ports - which we have already seen, emerged in October 1996.

The process of transformation in Poland's seaports also embraces the forwarding services market. The previously state-owned firm of C. Hartwig Gdansk has been privatised and transformed into a company with limited liability. The company's partners are such firms as the foreign trade centres of Stalexport, Centrostal, Ciech Chemicals Importers-Exporters and AtlanticTrans Ex from Katowice. Over and above these, the Ministry of Privatisation reserved 20% of the shares which, if regulations concerning privatisation do not undergo any changes, will be handed over to the staff of Hartwig (10% free and 10% purchaseable). C.Hartwig of Gdynia, on the other hand continues to operate as a state owned enterprise, but the general plans for privatisation are being worked out. These plans seem to indicate a trend towards transforming the firm into a workers' company with limited liability.

As in the port of Gdansk, C. Hartwig of Szczecin remains a state owned enterprise in which steps are being taken to transform it into a company with limited liability with the participation of the employees.

Alongside the Polish-Czech-Slovakian forwarding company "Spedrapid" in 1992, the Polish-Byelorussian Forwarding Trading Company "Mirtrans" was established to handle transit traffic to and from Byelorussia and other countries of the Commonwealth of Independent States countries.

Alongside these port-maritime forwarding enterprises, many new forwarding firms have emerged, several constituting significant market players. Due to the present number of such firms, competition between them has intensified.

Considerable changes have also taken place in respect of shipbroking in Poland's seaports. In the ports on the eastern margin, for four years now, Morska Agencja Gdynia (MAG) has been a typical employees' company with limited liability. In Gdynia and Gdansk, there are now over 20 private shipbroking firms, which has resulted in stronger competition. Despite this, MAG has retained its dominant position, holding 50% of the ship broking agency market.

An analogous situation has arisen in the Szczecin Swinoujscie port group, where Morska Agencja Szczecin (an employees' company with limited liability) has been privatised. In the same port group, a dozen or so new brokerage firms have started up.

In the field of experts and cargo control services, privatisation is advancing. One of the largest enterprises in this branch, Polcargo, broke up into a dozen or so smaller companies. Alongside "Ship control" expertise and control enterprise which was converted into a limited liability company, and Supervise, a company with Polish-Swiss capital, J.S. Hamilton Poland Ltd, expertise and control firm, has existed for a number of years and is one of the three largest and most dynamically developing firms of its type. As from 1 January 1993, this firm has also been the owner of the limited liability company "Polcargo" Central Chemical Laboratory, the services of which are used by other firms of the branch. Apart from the services mentioned in this sphere, about 40 firms sprang up in Polish ports, 20 in Gdynia alone, but as they do not belong to the international control organisations, their certificates are not recognised internationally.

In the field of ship chandling in Polish seaports, "Baltona" state owned enterprises was transformed into a limited liability company, with LOT airlines holding a substantial number of shares. The difficult financial situation at Baltona, opened the way for new firms to step into the ship chandling business in Poland's ports. There are now over 20 such firms, which has substantially increased the competition in this field of port services.

The processes of restructuring and privatisation of service enterprises in Poland's seaports are gradually taking place and are moving in the direction of creating an effective port services market. Such processes which are taking place in the existing state owned enterprises should be completed by the end of 1997.

Printed and bound by CPI Group (UK) Ltd, Croydon, CR0 4YY

21/10/2024

01777084-0003